R. A. MANGUSHEV,  A. V. ERSHOV

# PILE CONSTRUCTION TECHNOLOGY

Second Revised and Updated Edition

ASV Construction
Stockholm, Sweden
2016

Universal Decimal Classification: 624.154

*Reviewers*:
Head of Construction Operations Technology Department
Saint Petersburg State University of Architecture and Civil Engineering,
Doctor of Engineering Science, Professor *V. V. Verstov*;

Geotechnics Department, Saint Petersburg State University of
Architecture and Civil Engineering, Merited Scientist of Russian
Federation, Doctor of Engineering Science, Professor *A. B. Fadeyev*

**R. A. Mangushev et al.**
Pile Construction Technology / R. A. Mangushev, A. V. Ershov,
A. I. Osokin; 2nd Edition, revised and updated, Stockholm, Sweden,
ASV Consruction, 2016. – 228 pages.

ISBN 978-91-982223-0-2

The pile construction technologies are considered in the context of the compact urban development in case of new construction and reconstruction. Their advantages and disadvantages are set forth. The pile penetration technology using the jacking-down method is discussed. Special consideration is given to the modern technology of constructing replacement and displacement piles. Besides, screw steel piles are discussed that recently have been brought into active use in production and civil construction including the construction of foundations for country and low-rise houses.

The last chapter is dedicated to the quality control of piling works.

The Appendices to this Manual include the technical features of rigs and accessories for the penetration of driven, jacked, screw, replacement and displacement piles.

This Manual is intended for the students of civil construction institutions of higher education who study building disciplines, students of upgrading qualification institutes and engineers and technicians who specialize in geotechnical engineering.

ISBN 978-91-982223-0-2

# CONTENTS

# FOREWORD

Pile foundations are widely used in the domestic and foreign practice of foundation construction. They allow constructing buildings and structures on soft soil with insufficient bearing capacity, and in many cases, it is the only construction method in difficult engineering and geological conditions.

The major advantages of pile foundations are the reduction in construction time, the high adaptability to construction process and the reduced labor content and earthwork volume.

Before 1990s, in the Soviet Union, prefabricated piles (in 70 % of applications) prevailed that were sunk into soil using various methods, among which the most popular was the driving method. Later on, the modern low-impact methods of pile penetration started to be used increasingly often, in particular, the pile-jacking method for construction in areas of dense urban development.

From the middle of 1990s, piles constructed in soil (replacement and displacement piles) started to be widely used along with precast piles. The piles constructed in soil are used more and more frequently when high concentrated loads are transferred to the foundations of buildings and structures to be constructed on soft soil, as well as when foundations are constructed in the vicinity of the existing buildings and when reinforcing the foundations of operated facilities, etc.

In industrially developed countries, an extensive experience has been accumulated for the high-quality construction of foundations using replacement and displacement piles. The most advanced technologies are currently used in Russia and, in particular, in Saint Petersburg.

The first edition of the "Modern Pile Construction Technology" was printed by the Publishing House of the Saint Petersburg State Architectural and Construction University (first issues) in 2007 and by the Publishing House of Association of Civil Construction Universities in Moscow (second issue).

This edition of the Manual is the second one. It was revised and updated with the materials that were not included in the previous edition

The second edition gives a more detailed review of the modern technologies for the construction of pile foundations. The construction methods are described for the piles constructed in soil that were prototypes of modern pile construction technologies. The chapter dedicated to the quality control of piling works was added. Operational experience of the most advanced Saint Petersburg geotechnical companies was reviewed for the last 20 years.

In an additional appendix, the technical features are given for a number of modern rigs and machinery used for pile penetration. This data can be useful for the diploma students and engineers who specialize in development of Plan for Organization of Construction (POS) and Program of Work (PW).

This Manual is intended for the students of building disciplines, students of upgrading qualification institutes who specialize as construction engineers, and engineers of designing and construction companies.

The authors express their gratitude to the manuscript reviewers: Professor A. B. Fadeyev, Doctor of Engineering Science, Merited Scientist of Russian Federation, and Professor V. V. Verstov, Doctor of Engineering Science, whose comments were taken into consideration in preparing this Manual for the publication.

In many ways, the initiative and assistance in the publication of the second edition of this book have been provided by N. V. Balberova, Ph. D., Director of the Upgrading Qualification and Occupational Retraining Institute under the Saint Petersburg State University of Architecture and Civil Engineering. The authors greatly appreciate this effort and express their sincere gratitude to Dr. Balberova.

Please send your comments and recommendations related to this Manual to the following address: 4, 2nd Krasnoarmeyskaya Street, Saint Petersburg 190005 Geotechnics Department, Saint Petersburg State University of Architecture and Civil Engineering. E-mail: geotechnica@spbgasu.ru, andrew_ershov@mail.ru

# INTRODUCTION

Currently, the construction market in Saint Petersburg is represented by a large number of companies specializing in foundation construction of various types including also pile foundations. Many different technologies are offered with different quality and prices of works to be performed. Accordingly, as often as not, a customer faces a difficult choice of the most efficient and safest pile technology and construction. This choice should be made depending on the specific geological and hydrogeological conditions, the design of a building (structure) to be constructed or reconstructed, the adjacent buildings and their technical conditions.

The substantial assistance in making such choice shall be provided to a developer by specialists – geotechnical engineers - at all stages of the designing and design support. It is these specialists who can make the correct assessment of the engineering and geological conditions of a construction site (stratification, type and properties of soil), offer the most efficient foundation design for a projected facility with due consideration of its sensibility to non-uniform settlements and recommend the most effective technologies (including technologies of low impact on adjacent buildings) of construction operations.

Saint Petersburg is the first of the Russian cities where the advanced soil improvement methods and foundation construction technologies, in the first place, pile foundations, started to be used. To a large extent, the prerequisites were the difficult engineering and geological conditions of the city location where numerous glaciers passed and seas and glacial lakes appeared in different geologic periods.

In the central part of the city, the relatively firm moraine deposits occur at a depth of 15 to 30 m from the ground surface. When performing the zero-cycle works in Saint Petersburg, it is required to take into consideration the soil properties of the above-moraine layer represented by late-glacial and post-glacial lacustrine and marine deposits. It is these soils that very often serve as a base of shallow foundations, and the major part of the friciton pile shaft of the pile foundations is located in these soils.

When using such soils as a base of foundation, the following occurs:
- large, non-uniform, long sustained settlement of buildings, structures and adjacent area;
- loss of stability in bearing layers of building and structure foundation bases composed of clay soils in the conditions of non-completed consolidation or exposed to freeze-thaw action;
- destruction of natural soil structure when performing earthworks using the conventional methods;

- occurrence of quicksand in case of surface drainage from founda-tion pits and trenches;
- change in the bearing capacity of piles due to the development of neg-ative friction force in the areas elevated with filled or hydraulically filled soil;
- development of putrefaction of peat, organic impurities in soil and timber elements when groundwater recession occurs.

Many construction sites are characterized by the large thickness of soft water-saturated thixotropic soils, their considerable non-uniformity in plan view and by depth both in terms of composition as in terms of physical and mechanical properties.

The allowance for these complex processes determines, in many in-stances, the professionalism and success of construction companies engaged in operations related to the construction of bases and foundations.

For the recent years, the experience of the construction and recon-struction in the central part of Saint Petersburg shows that the buildings of the adjacent development are exposed to higher deformations during the zero-cycle works than during the static loading of the foundations of new buildings or during the additional loading of reconstructed buildings.

To provide the normal operation of the adjacent buildings, recon-structed and constructed buildings are increasingly often designed on pile foundations. In the central part of Saint Petersburg, replacement and dis-placement piles 20 to 35 m long are used in order to cut through soils of the above-moraine depth and to transfer a load to underlaying low-compressible soil. The use of the piles reduces the self-deformation of the reconstructed or newly constructed buildings and the additional deforma-tion of the adjacent buildings (Figure I.1).

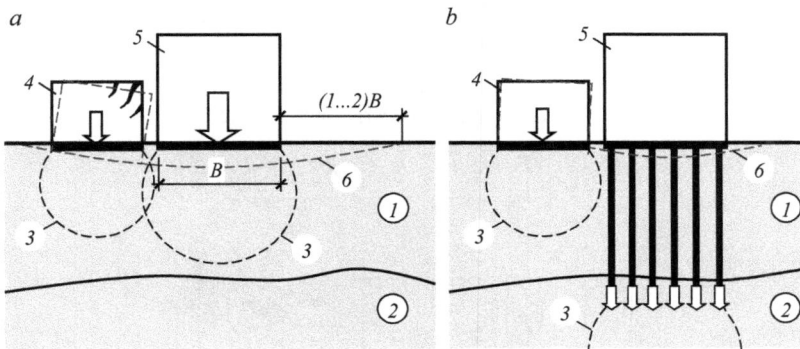

Figure I.1. Models of impact on existing building from new building constructed on raft foundation (*a*) and pile foundation (*b*): *1* – medium-compressible soil; *2* – low--compressible soil; *3* – stress-development area; *4* – existing building; *5* – building under construction; *6* – surface subsidence crater resulted from foundation loading of new building

Figure I.2. Operation of Junttan Rig
when driving reinforced concrete piles

Figure I.3. Operation of Banut 655 Rig
when driving reinforced concrete piles

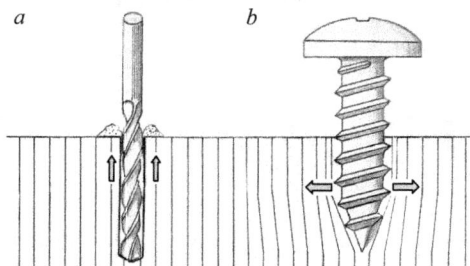

Figure I.4. Methods of borehole-making for pile installation:
*a* – drilling (soil removal); *b* – reaming (soil displacement)

For the last 15 years, the city saw the significant change in the approach to zero-cycle works. The construction companies specializing in geotechnical works came to the construction market, and the advanced equipment and technologies appeared and are widely introduced.

For example, the use of driven piles became half as much as compared to 1990. The driven piles have been replaced by more advanced low-impact technologies: jacking down of piles (including the penetration into pilot boreholes), use of replacement and displacement piles of various lengths and diameters, method of diaphragm wall, etc.

Until the beginning of 1990, the basic type pile foundation were the foundation of driven reinforced concrete piles, the use of which has proved to be extremely dangerous for old buildings when new buildings are constructed in close vicinity to the old buildings. Many failures and damage to the adjacent buildings occurred when piles were driven at a distance up to 20 m (and in some instances, more than that), and designers and builders had to search for and implement the technologies of lower impact. When driving the piles, Junttan (Figure I.2) and Banut (Figure I.3) hydraulic hammers with the hammer strike frequency up to 100 strikes per minute and with a weight of the striking part of 30 to 250 kN are used very often.

The partial abandonment of the driving method resulted in the development of the pile jacking technology and in the implementation of new technologies related to the construction of compaction piles. The above technologies can be divided into three groups: pile penetration with soil removal, without soil removal and with partial soil removal.

The first group includes *replacement piles* concreted in drilled boreholes, from which soil is removed to the surface (Figure I.4). The second group includes *displacement piles* constructed in boreholes that are made by way of soil displacement when a borehole is penetrated by displacement tools, closed-end tubes and other special-purpose tools. *The piles installed with partial soil displacement (small displacement piles)* included in the third group are constructed by a technology, which uses the partial removal and forced displacement of soil.

Widely used are replacement piles constructed using the continuous flight auger with or without a casing pipe, Titan and Franki bored injected piles, Atlas, Bauer, Tubex and Fundex displacement piles. In some cases, the pulse discharge technology is used for the installation of piles.

To construct pile foundations of replacement and displacement piles, the rotary drilling rigs are used that are manufactured by Bauer, Casagrande, Modern Drilling Equipment, SoilMec, Stroydormash. The multifunctional rigs made by Liebherr, RTG Rammtechnik, Fundex, Junttan, etc. also found common use. The technical features of the above rigs are given in the Appendices.

# Chapter 1. PREFABRICATED PILES

## 1.1. Jacked piles

In the middle of 1970s, Leningrad became one of the first cities where the jacking technology for prefabricated piles started to be used on an industrial scale. A non-self-propelled rig designed by Trust No. 101 of Glavleningradstroy (Figure 1.1) allowed jacking the piles with a cross section of 35×35 and 40×40 cm, up to 32 m long at the maximum force of 2000 kN.

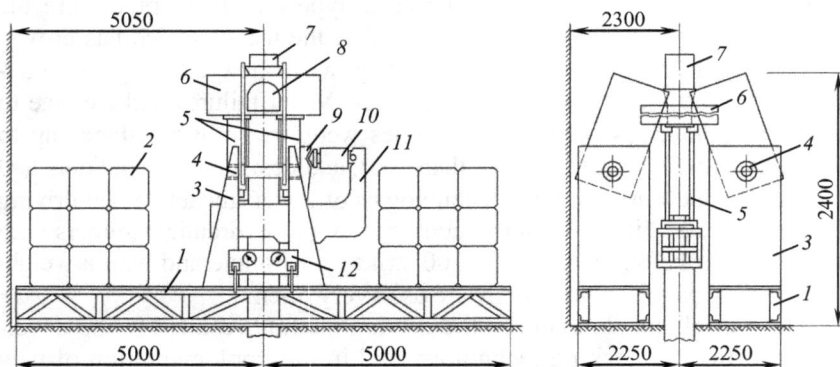

Figure 1.1. Non-self-propelled pile-jacking rig designed by Trust No. 101:
*1* – three-dimensional beam of loading platform; *2* – removable weight;
*3* – bracket; *4* – pin; *5* – hydraulic jack for pile jacking; *6, 11* – upper and lower part, respectively, of working tool of rig; *7* – pile; *8* – oil-pumping station;
*9* – arm; *10* – hydraulic jack for pile clamping; *12* – control panel

The base of the rig is a loading platform made of two three-dimensional lattice beams to accommodate the weight. On the beams, brackets were installed to secure a hydraulic working tool of the rig by means of pins.

The working tool in the form of a cross traverse consisted of an upper and lower beams, on which the pile clamping and jacking mechanisms were installed.

In the upper and lower beams, there were provided with openings, through which a pile passed. To clamp the pile, a hydraulic jack was used that was placed horizontally on the lower beam. The hydraulic jack transferred the clamping force through the arm to a movable plate (cheek), which pressed down a pile to the wall of the opening in the lower beam.

The pile was jacked down with two hydraulic jacks. Their hydraulic cylinders were mounted vertically and in parallel to each other on the upper beam, and the rods were connected by means of the lower beam. Five-tonne cast iron blocks were used as weight (total weight up to 170 t).

The clamping and jacking mechanisms were operated by two oil-pumping stations powered by electric motors. The rig was controlled from the control panel, which accommodated hydraulic control valves with operating levers of the pile clamping and jacking mechanisms, as well the pressure gages to control the clamping and jacking forces.

All handling operations to mount/dismantle the rig and to load the piles were performed by means of a crane with the lifting capacity of at least 10 t. The lifting height of the crane hook was selected depending on a length of the piles to be jacked.

The rig was serviced by a crew of three people: rig operator, crane operator and rigger. The rig operator was in charge of the pile loading, operated the rig during the pile jacking, and together with the rigger dismantled and mounted the rig when it was moved from one location to another. The rigger took part in the handling operations to load the piles and to mount and dismantle the rig, as well as adjusted the actions of the rig operator and crane-operator.

Trust No. 101 used the following technology for jacking the piles by means of their rig. The rig was installed at a pile penetration point. A pile was inserted vertically into the working tool of the rig and clamped by the clamping mechanism. The pile was driven to a length of the piston stroke of the tandem hydraulic jacks of the jacking mechanism (down to 1.0 m). Then the pile was released in the working tool of the rig and the tandem hydraulic jack rods of the jacking mechanism were lifted to the upper position. After that, the pile was again clamped, and the cycle was repeated from the beginning. After the pile has been driven to the design depth, the rig was dismantled and moved to the next penetration point.

The rig was easy-to-operate, not power-consuming and allowed driving the piles of the bearing capacity up to 2000...2500 kN into clay soils with the index of liquidity $I_L > 0.2$.

The disadvantages of this rig were the low efficiency due to the required dismantling for each pile, the absence of the direction adjustment in the process of the penetration and the large overall dimensions that did not allow using the rig in the conditions of compact urban development.

The necessity to perform construction operations in the confined area of compact urban development (including the historical part of the city) resulted in the fact that, from the middle of 1980s, the Glavleningradstroy Trust (now ZAO Construction Trust No. 28) began to use the USV-80 self-propelled pile-jacking rig mounted on the EO-6122 crawler excavator designed by VNIIGS, Gersevanov NIIOSP and other agencies.

The USV-120 rig (Figure 1.2) includes a base machine and a weight trolley and allows jacking the piles with the face size 30 to 40 cm, 4 to 16 m long at the maximum force of 850 kN at a distance of 1 m from an existing building. At a distance of 4.5 m from a building, the rig can jack piles of the same length with the cross sections of 30×30 cm and 35×35 cm at a force up to 1200 kN, and with the cross sections of 40×40 cm at a force up to 1800 kN. In the soil conditions of Saint Petersburg, the rig capacity is eight to two precast sectional piles (16 to 28 m long, respectively) per shift.

The rig is provided with split mast composed of three sections. The mast height is to be selected depending on the pile (pile section) length and the confined conditions in terms of height. The angle of lean in four directions is controlled by the Vertical-20B instrument.

The rig is serviced by a crew of four people: USV-120 rig operator, crane operator and two riggers. The rig can be spot-turned by means of jacking it on the outriggers.

The USV-120M modernized rig (Figure 1.3) has been successfully used in Saint Petersburg more than 15 years; its weight is greater (117 t instead of the previous 105 t) and it is provided with a boom and a drilling mechanism and a leverage mechanism. The USV-120M rig is designed for the use in the clay soil conditions with the index of liquidity $I_L > 0.3$.

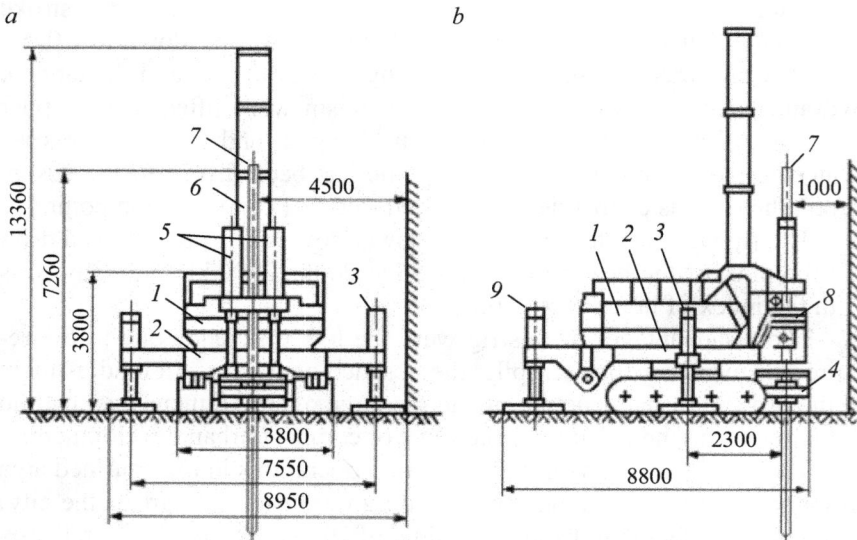

Figure 1.2. Pile-jacking rig USV-120: $a$ – front view; $b$ – side view;
$1$ – undercarriage of excavator EO-6122; $2$ – underframe; $3$ – side outrigger;
$4$ – clamping mechanism; $5$ – pile-jacking hydraulic cylinder; $6$ – mast;
$7$ – pile; $8$ – weight; $9$ – rear outrigger

12

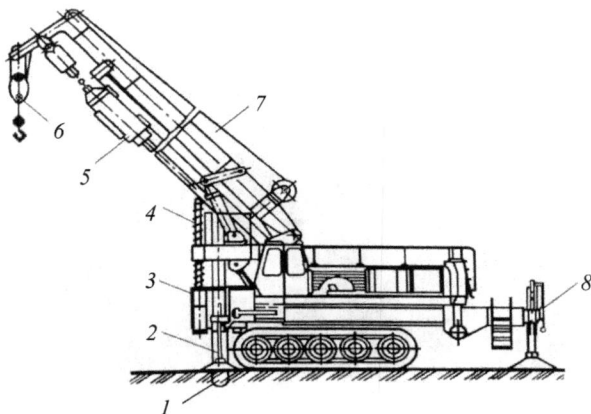

Figure 1.3. Pile-jacking rig USV-120M:
*1* – pile; *2* – side outrigger; *3* – clamping and jacking mechanism;
*4* – auger; *5* – auger head; *6* – hook block; *7* – boom; *8* – rear outrigger

The boom with a hook block can be installed on the front part of the underframe to clamp and deliver the piles to the rig working tool, as well as to perform other load-handling operations. Due to the installation of the boom, no auxiliary crane is required. Besides, the hook block can be connected to an anchor to partially take up the reactive force.

A drilling mechanism with a set of augers is used to drill pilot boreholes up to 300 mm in diameter and up to 21 m deep.

The leverage mechanism of the rig is designed to clamp the sheet piling and allows driving and extracting the Larsen-IV и Larsen-V steel sheet piling.

Currently, the Trust has the pile-jacking rigs operating at the jacking force up to 1600 kN (Figure 1.4), the design is finalized for a rig operating at a force up to 2000 kN and a bridge-type rig is being designed that is provided with the end transfer of the force to pile.

The SVU-V-6 pile-jacking rig is mounted on the RDK-250 crawler crane (Figure 1.5). It is designed to jack square and rectangular piles with the cross section up to 35×35 cm, circular piles up to 35 cm in diameter and sheet piling of any type. The rig allows driving piles below the soil surface.

The attachments allow jacking piles at an angle of 20 degrees from the rig and below the rig. The minimum distance is 0.4 m from the centerline of the jacked pile to the wall of the existing building.

The jacking is done by means of sheave-block mechanism at the maximum force of 900 kN. The rope-and-block system of the rig allows continuously driving piles and, at the same time, controlling the driving process.

Figure 1.4. Pile jacking with USV-160 rig

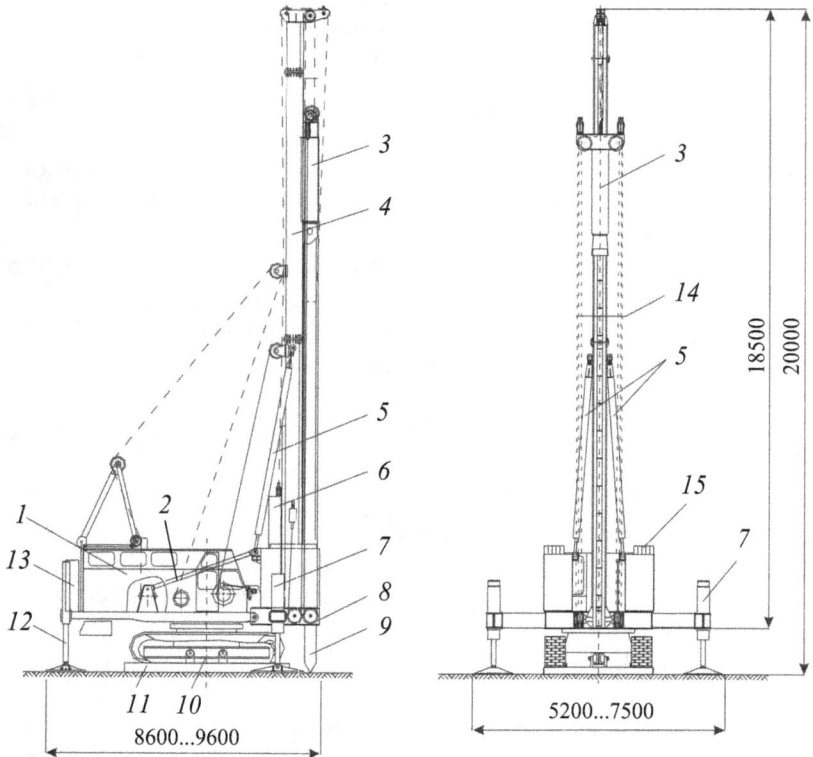

Figure 1.5. Pile-jacking rig SVU-V-6: *1* – basic machine; *2* – pull rod; *3* – working tool; *4* – pile-driving support; *5* – cross stay; *6* – counterweight; *7* – side outrigger; *8* – supporting frame; *9* – pile; *10* – thrust beam; *11* – supporting slab; *12* – rear outrigger; *13* – hydraulic unit; *14* – pressing-in sheave block; *15* – weight

14

The SVU-V-6 rig performs all jacking-related technological operations. The rig makes it possible to unload piles, to drag them to the installation area and to install them at the penetration point. The rig in self-propelled mode travels across the site, drives onto means of transportation and drives down from them. The unladen weight of the rig is 112 t, and the transport weight is 59 t.

The use of the supporting slab provides the uniform distribution of load from the rig weight and the significant reduction in the pressure on the base. At the required jacking force up to 400 kN, the rig can be used without the supporting slab and weights.

In 2008, the Sunward walking-type pile drivers made appearance on the Saint Petersburg construction sites. The Sunward pile drivers are designed for jacking square piles, rolled steel sections, pipes and sheet piling. The Sunward pile drivers include 13 models with their maximum jacking force ranging from 800 to 12000 kN (see Appendix 1).

The design of the pile drivers is a rectangular platform, in the middle of which the main hydraulic jacking mechanism is mounted (Figure 1.6). The side jacking mechanism and the cabin of the pile driver operator, who is in charge of the pile jacking and the pile driver traveling, is situated at the one side of the platform, and the turnplate with the telescopic crane boom and the crane operator's cabin is located on the opposite side.

Figure 1.6. Sunward pile driver:
1 – pile driver platform; 2 – turnplate with boom and crane-operator's cabin; 3 – telescopic boom; 4 – central jacking mechanism; 5 – pile driver's cabin; 6 – side jacking mechanism; 7 – side outrigger; 8 – supporting beam with longitudinal guide rail; 9 – supporting slab with transverse guide rail

The pile driver platform bears on the site surface by means of the supporting slab and/or supporting beams. Between the supporting slab and the platform, there is a turning gear, which allows the slab and platform to turn relative to each other.

The Sunward pile drivers from ZYJ 420 to ZYJ 1200 are completed with the side jacking mechanism of the mounted-type only, which allows generating a force of no more than 40 % of the rated value.

The Sunward pile drivers travel using the supporting slab and two supporting beams mounted on the side outriggers. On the supporting beams, there are fastened the guide rails, along which the pile driver travels longitudinally. The similar guide rails are mounted on the supporting slab transversely the pile driver platform.

To move the pile driver longitudinally, the platform together with the supporting slab is raised by the outriggers above the site surface and travels along the guide rails on the wheel pairs that are fastened at the ends of the supporting beam outriggers. Then the supporting slab together with the platform is lowered onto the site surface. For the further longitudinal travel, the supporting beams are raised by means of the side outriggers, moved in the required direction and lowered down. That is how the longitudinal pace of the pile driver is carried out.

The pile driver travels longitudinally relative to the supporting beams and transversely relative to the supporting slab. For this purpose, the supporting beams are raised by means of the outriggers, and the pile driver platform travels along the guide rails of the supporting slab. Then the platform together with the supporting slab are raised by means of the side outriggers, and the supporting slab travels relative to the platform and lowered to the surface. That is how the transverse pace of the pile driver is carried out.

The Sunward pile drivers have large dimensions in plan view, which makes it difficult to use them on small construction sites. To provide the unobstructed travel of the pile driver, it is required to perform leveling operations and to strictly comply with the process sequence of the pile driving. Otherwise, problems might arise to move the pile driver to omitted piles. The pile driving time is not long. It takes considerable time to position the pile driver, to align the platform, to attach hoisting slings and to bring a pile into the clamping mechanism.

The technical features of the pile drivers concerned are given in Appendix 1.

When selecting a pile driver for the static jacking, one should take into consideration the pile design, the required jacking force and the confined conditions of the construction site in terms of square area and height.

To drive a pile, the jacking force shall be higher than the bearing capacity of the pile ground base $F_d$ determined by the calculation method according to SNiP 2.02.03 or SP 50-101: 1.4 times for the clay soil and 1.9 times for sandy soil.

During the pile driving, it is required to control the clamping and jacking forces, the vertical position and depth of the pile penetration.

To control the clamping and jacking forces in the hydraulic pile jacking rigs, pressure gauges are used that measure the side and axial load on a driven pile against the oil pressure in the clamping and jacking hydraulic cylinders, respectively.

The modern pile-jacking rigs are provided with on-board computers or printers that allow continuously monitoring the jacking force and the pile penetration depth.

At the final stage of the pile penetration, the jacking force should be measured the most thoroughly (operational control) in order to assess, to the adequate degree of accuracy, the actual pile bearing capacity $F_d$ by Formula (4.14). Based on the pile jacking data, a particular value of the pile ultimate strength $F_u$ (kN) can be approximately estimated by the following formula

$$F_u = F k_v k_t,$$ (1.1)

where $F$ = jacking force along the last 10 cm of pile penetration, kN; $k_v$ = factor allowing for the effect of pile penetration speed on jacking force; at the jacking speed up to 3 m/min, it should be taken as $k_v = 1$; $k_t$ = pile "suction" factor.

For clay soil

$$k_t = 1 + b I_p I_L,$$ (1.2)

where $b$ = empirical factor taken as equal to 16.5; $I_p$, $I_L$ = pile-length weighted average values of plasticity index and index of liquidity, respectively, unit fraction.

For sandy soil

$$k_t = 1 - \frac{c}{K_\phi},$$ (1.3)

where $c$ = empirical factor taken as equal to 0.02; $K_\phi$ = coefficient of permeability, m/day.

The advantages of the pile jacking technology are the following:
- high quality of the pile shaft and the penetration accuracy due to the use of prefabricated piles;
- insignificant (as compared to driven piles) dynamic loads both on a pile and on adjacent buildings and structures;
- low noise level due to the use of the rigs with the electrohydraulic drive;
- possibility to continuously monitor the jacking force and, therefore, to estimate the bearing capacity of a pile driven.

Among the disadvantages of this methods are the following:
- it is difficult to create a high jacking force in terms of magnitude;
- it is required to build temporary roads of reinforced concrete slabs to make it possible for the rig with the tremendous weight to travel across the construction site;

- limited application of this method in certain soil conditions, in particular, in the soil layers containing major solid inclusions, in compact and medium-compact sand, as well as in clay soil of hard and semi-hard consistency;
- it is required to provide the construction site with a power supply source of considerable capacity (up to 200 kW), which requires, in the site conditions, the use of the independent power supply (diesel generator);
- possibility of some dynamic impact on the surrounding soil mass when a pile penetrates through compact soil.

In soil with laminated structure, the pile penetration might result in additional deformation in the bases and foundations of the buildings located at a distance up to 3 m.

To reduce the pile driving resistance of soil, there is used the water-jet method of pile driving method, the preliminary soil ripping method or the method of drilling pilot boreholes with a diameter smaller than the size of piles to be driven. In particular, one of the above methods was use on the construction site at 50, Shpalernaya Street when 28-meter piles with a cross section of 35 cm were driven near the existing buildings that had cracks in the enclosing structures. The soil conditions of the site were characterized as unfavorable due to the high compressibility and thixotropy of soil. In silty sand loam and loam, there occurred interlayers of gravel sand with boulders and medium-hard sand loam.

When driving piles by means of the Glavleningradstroy's non-self-propelled rig, the acceleration rate of the ground vibration reached $0.16$ m/s$^2$, which is higher than the maximum permissible value of $0.15$ m/s$^2$ according to VSN 490-87 (the monitoring was carried out by ZAO Georekonstruktsiya-Fundamentproekt). The dynamic effect on soil was related to the fact that when the soil reactive resistance increased up to 1400 kN and higher, the rig hit the ground with all its weight at the time of engaging a pile with the clamping device [2].

In view of that, the further pile penetration was carried out by means of the USV-120M rig with ripping soil by an auger 300 mm in diameter to a depth of 24 m. That allowed reducing the maximum vibration acceleration rate of the adjacent building wall down to $0.055$ m/s$^2$. The use of the additional weight in the form of reinforced concrete slabs that were placed on the brackets mounted on the rig frame resulted in even more significant reduction in the rate of vibration acceleration: down to $0.015$ m/s$^2$.

When determining a method to reduce the pile jacking resistance of soil (pilot-borehole drilling, preliminary ripping by auger, water-jet method), it is required to take into consideration both the possible reduction in the pile bearing capacity and the negative effect of the above measures on the foundation conditions in the existing buildings and structures.

For example, the dynamic loads prove to be comparable with the loads induced by the pile penetration in case of the preliminary ripping of the compact soil layers, especially the gravel interlayers with boulders. The partial removal of soil from boreholes might result in the soil movements at the penetration level, which, based on the SPbGASU observations, affects the settlement of adjacent buildings.

*Precast sectional piles* are used when a thick layer of soft soil occurs at the foundation construction location.

Elements (links, sections) of precast sectional piles with the cross section of 30×30, 35×35 and 40×40 cm have a length up to 16 m and are connected, at the time of penetration, by means of welding, bolting and special-purpose butt joints or locks.

There are several designs of pile joints. The *welded spliced joint* (Figure 1.7) is the most commonly encountered on construction sites of Saint Petersburg.

When jointing, a lower element of the sectional pile is driven to a depth in such a manner that it protrudes 1.0 m above the ground level. The next element is placed on the pile head of the driven element, and a joint is made by means of splice pieces of steel sheet that are welded to the embedded parts of the pile elements. It might take 40 min. to make the above joint.

*A bolted joint* (Figure 1.8) includes two metal retainers consisting of pressed parts of steel sheet. The parts have a flat end plate and a retainer, which contributes to taking up impact loads in driving. The design of the joint allows jointing the elements on any end face and disjointing them if required. The duration of bolting the pile links is 20...25 min.

*Sleeve-type joint* provides the automatic connection of the pile links in the pile penetration process and consists of a metal sleeve made of a circular- or rectangular-shaped pipe section. At the open end, the sleeve is beveled and rigidly fixed on the lower link of the pile by means of starter bars welded to the inner side of the pipe sleeve (Figure. 1.9). The next link is tightly set up in the steel sleeve, with the lower end of this link having a cylindrical reinforced concrete section provided at the free beveled edge. The time to joint the pile links is approximately 15 min.

*A joint with wedge-shaped dowels* includes two steel retainers made of welded angle pieces (Figure 1.10). Four rectangular-shaped holes are cut through the retainers on the end-face side. In the retainer of the upper element, metal loops are inserted in and welded to those holes. In the retainer of the lower element, there are chambers made of angle pieces of the same or somewhat smaller steel section. Four circular holes are drilled in the chamber walls on the side face.

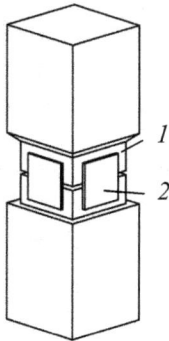

Figure 1.7. Welded spliced joint:
*1* – pile head; *2* – splice piece;

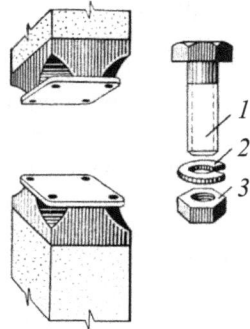

Figure 1.8. Bolted joint:
*1* – bolt; *2* – washer; *3* – nut

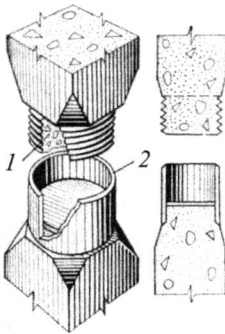

Figure 1.9. Sleeve-type joint:
*1* – lower corrugated concrete end
face of pile link; *2* – steel sleeve

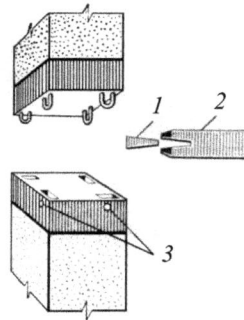

Figure 1.10. Joint with wedge-shaped
dowels: *1* – wedge; *2* – dowel;
*3* – hole to insert dowel

After the upper element has been driven, the lower element is placed on it in such a manner that its protruding loops find their way into the rectangular holes. Then, four pins with wedges are driven in the side holes by a hammer.

In principle, a *pin-type joint* (Figure. 1.11) differs little from the joint with wedge-shaped dowels. The pile links have metal head caps, on whose ends there are two locking dowels and two sockets. The links are connected by way of inserting the dowels in the sockets. Then four steel pins are driven into holes made along the side surfaces of the head cap, with the pins wedging the dowels in the sockets. The duration of making the pin-type joint is 15 min. Table 1.1 shows the basic dimensions of the pin-type head caps.

20

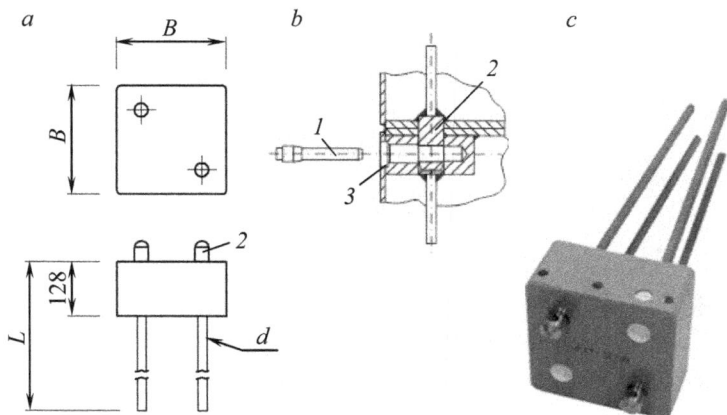

Figure 1.11. Pin-type pile joint: *a* – design of head cap; *b* – pin joint assembly;
*c* – general view of head cap; *1* – pin; *2* – locking dowel; *3* – socket

*Table 1.1*

**Range of Leimet pin-type head caps**

| Pile cross section, mm | $B$, mm | $L$, mm | $d$, mm |
|---|---|---|---|
| 235×235 | 232 | 700 | 16 |
| 250×250 | 247 | 700 | 16 |
| 270×270 | 267 | 770 | 20 |
| 300×300 | 297 | 770 | 20 |
| 350×350 | 347 | 770 | 20/25 |
| 400×400 | 397 | 850 | 25/32 |

In addition to the above jointing methods, there are also the follow-ing (Figure 1.12):
- free connection of pile links by means of a conductor installed on the head cap of the lower link;
- lock connection due to wedging gaps in the lower pile link with special-purpose throw-over locks;
- bonded key joint of shaped pile links. The upper part is made in the form of a projection, while the lower part is made in the form of a socket. After the sections have been installed, a gap is filled with epoxy resin adhesive;
- bonded dowel joint. Starter bars (dowels) of the upper link are in-serted in the lower link holes filled with adhesive;
- hinge joint of sleeve type using hardening colloid-cement or epoxy resin adhesive.

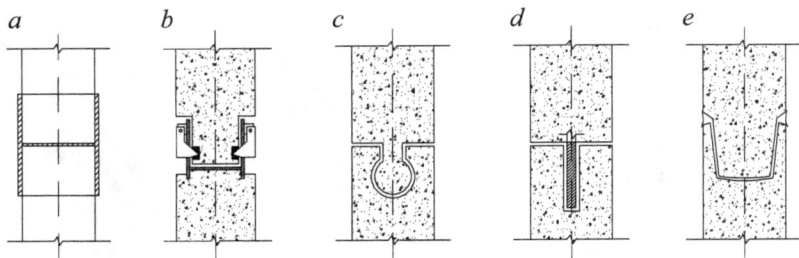

Figure 1.12. Pile joint designs: *a* – joint using conductor; *b* – locking joint; *c* – bonded key joint; *d* – bonded dowel joint; *e* – bonded sleeve-type joint

## 1.2. Screw piles

Screw piles consist of a steel pipe with a screw blade that provides the driving of piles by rotation.

Cast and welded screw blades are made of carbon (VSt3sp5, 09G2S) and low-alloy (10khSND, 10G2SKh) steel. The blade diameter shall not be greater than 4.5 times the pipe diameter and more than 3.0 m.

The optimum parameters of the screw points are as follows: screw blade spacing: 200...250 mm, shaft (hub) diameter: 168, 219, 278 and 325 mm, screw blade diameter: 500, 700, 850 and 1000 mm [6].

Due to screw blade, whose diameter is greater than the pipe diameter, and the high bearing capacity of steel, the screw piles can take up considerable pulling forces, which allows using these piles as anchors.

Another advantage of the screw piles is the absence, during piling operations, of significant dynamic and vibration impact on the structures of closely adjacent buildings.

A disadvantage of the screw piles, as any steel structures, is their proneness to corrosion, so the surface of the piles is treated with special-purpose coating to protect against corrosion. Zink-additive compound is most often used as primer, and abrasion-resistant compound is used as top coat.

A design of the screw piles depends on soil properties. In thawed soil, wide-blade screw piles are used (Figure 1.13, *a*) with the blade being of the variable width, which begins at the conical part of the screw point and smoothly transfers to the cylindrical part as the width increases. Such design provides the soil cutting with the minimum resistance; therefore, a smaller torque is required for the pile penetration. An angle and spacing of the screw blade allow screwing a pile without loosening soil and with the minimum axial weight.

In permanently frozen soil, narrow-blade screw piles are used (Figure 1.13, *b*) that are screwed into pilot boreholes, the diameter of which is equal to the pile shaft diameter.

Table 1.2 shows the basic design parameters of the screw piles developed by Sevzapenergosetproekt.

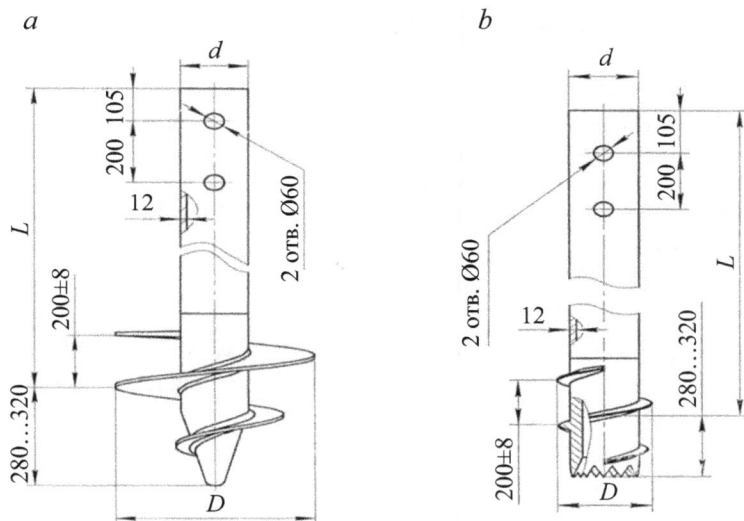

Figure 1.13. Steel screw piles with cast point developed by Sevzapenergosetproekt:
*a* – for thawed soil; *b* – for permanently frozen soil

*Table 1.2*

**Range of screw piles**

| Pile model | Pile length $L$, mm | Shaft diameter $d$, mm | Blade diameter $D$, mm | Blade bearing area, m$^2$ |
|---|---|---|---|---|
| SVL-15 | 5000±25 | 168 | 500±9 | 0.162 |
| SVL-25 | 5000±25 | 219 | 500±9 | 0.162 |
| SVL-28 | 5000±25 | 219 | 850±9 | 0.532 |
| SVLM-23 | 5000±25 | 219 | 300±8 | 0.037 |
| SVL-15-01 | 6000±31 | 168 | 500±9 | 0.162 |
| SVL-25-01 | 6000±31 | 219 | 500±9 | 0.162 |
| SVL-28-01 | 6000±31 | 219 | 850±9 | 0.532 |
| SVLM-23-01 | 6000±31 | 219 | 300±8 | 0.037 |

* The pile material is Steel S255 or S345 (09G2S).

With small scope of works, screw piles can be driven by means of a rotary mechanism, in the capacity of which a wheel can be used that is fastened to the pile head. One end of a rope is reeled on the wheel, another end is wound on the machine frame. When the machine moves, the rope reels out from the wheel and makes the pile rotate. The rotary mechanism allows screwing piles in cases when heavy-duty special-purpose equipment cannot be used (for example, when operating on ice or from a pontoon).

Small-diameter piles can be screwed manually, by means of a capstan (crow-bar). The capstan passes through the holes in the pile shaft and is rotated by two to four people.

In case of large-scale penetration of screw piles, the most efficient is the use of construction machines (excavators, drill-and-crane machines) with mounted hydraulic capstans – special-purpose screwing-in mechanisms (Figure 1.14). The hydraulic capstans allow screwing piles both vertically and at an angle.

The sequence of pile-screwing works using excavators includes:

- replacing the excavator bucket with the hydraulic capstan that has a torque of 100 kN·m;
- bringing the pile into a ratchet coupling and setting up the pile at the screwing location in vertical position or at the specified angle;
- screwing the pile to the design depth.

Figure 1.14. Screwing pile with hydraulic capstan
mounted on E-18 excavator boom

The bearing capacity is determined by the torque and controlled against the pressure in the hydraulic system using a pressure gauge. The required time is 30…40 min to screw`a pile to a depth of 5.0 m.

When constructing the Shuvalov's Village in Strelna, more than 500 piles with the shaft diameter of 219 mm and the blade diameter of 1000 mm were screwed by means of the hydraulic capstans mounted on the booms of the E-14 (E-18) wheel-type excavators.

The most proper rig to drive screw piles is a UBM-85 (UBM-150) general-service drilling machine mounted on the Ural-4320 truck. This machine can be used to drill boreholes with continuous flight auger 360 to 2000 mm in diameter.

The working equipment of the machine includes a string of pipes secured on a supporting rotary table and a three- or four-section telescopic boom. The supporting rotary table is connected to the frame of the base truck. A boom (a handling arm) is hinged to the string of pipes placed on the table. A rotary head with a reduction gear are located on the head cap of the telescopic boom with the swing radius of 1.8...12.0 m, which allows screwing several piles at one stop. The maximum theoretical length of a pile driven is 11 m. The boom is designed with consideration for the reactive torque generated at the rotary head when the drilling and screwing operations are performed. The machine is provided with a cargo platform, which allows transporting six screw piles.

Due to its overall dimensions, the machine can travel on public roads without restraint.

The pile screwing by means of UBM-85 is carried out in the following sequence:

- positioning of the machine at the pile-screwing location;
- stabilizing of the machine by means of four hydraulic outriggers (outboard supports) that are rigidly connected to the truck frame and create a support contour with the dimensions of 4.7×5.6 m;
- suspending of a pile in the vertical position by means of a winch mounted on the boom (Figure 1.15);
- fixing of the pile in the vertical position by means of compression in vice clamps mounted on the truck frame;
- securing of the pile with a wrench on the shank end of the output reduction gear;
- releasing of the pile from the vice clamps and positioning of the pile at the screwing location using the boom;
- screwing of the pile to the design depth (Figure 1.16).

The required time is 3...5 min to screw a pile to a depth of 5.0...6.0 m.

The technical features of the UBM-85, UBM-150 and MZS-219 machines for pile-screwing are given in Appendix 2.

Figure 1.15. Dragging of pile:
*1* – handling boom; *2* – hydraulic capstan;
*3* – chuck to secure pile; *4* – pile

Figure 1.16. Screwing of pile with hydraulic capstan MV-85
mounted on boom of UBM-85 machine

*a*

*b*

Figure 1.17. Examples of screw pile application in construction of
foundations for cellular transmission towers (*a*) and low-rise buildings (*b*)

Screw piles are used in the construction and reconstruction of power
transmission lines, contact-line supporting structures of railroads, cellular
transmission towers (see Figure 1.17, *a*), bridges. In addition, the screw
piles of small diameter (blade diameter = 300 mm, shaft diameter =
108 mm) find their application in the construction of low-rise country
houses (see Figure 1.17, *b*).

# Chapter 2. PILES CONSTRUCTED IN SOIL

## 2.1. Bored piles

Augers, drill buckets, rotary core bits and reamers (Figure 2.1) are used to drill boreholes for bored piles. Drilling tools are secured at the end of a drilling rod and driven into soil by means of a mechanism that transfers the torque and pressing force.

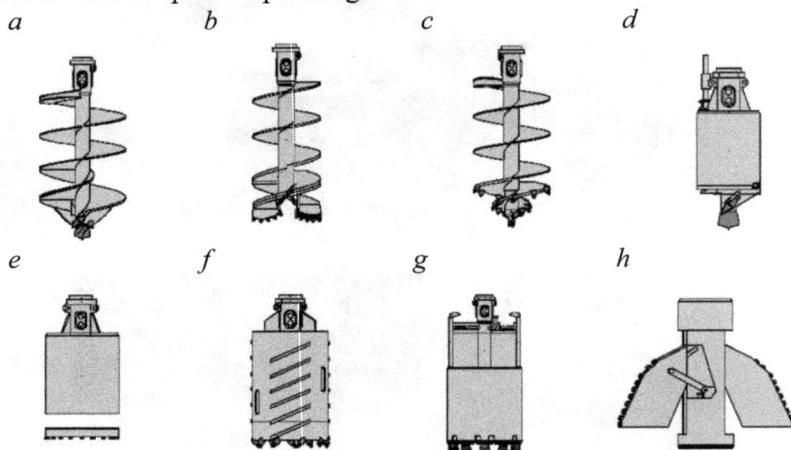

Figure 2.1. Tools to drill boreholes:
*a, b, c* – short-length augers; *d* – drill bucket;
*e, f, g* – rotary core bits; *h* – borehole reamer

Bucket augers (drill buckets) are used to clean the borehole bottom and to drill in sand, silt and clay.

The rotary core bits are used to drill out bridges along the pile length and to penetrate into rocky soil. The need for the rock excavation arises in constructing end-bearing pile, whose foot shall be buried in rock at least by 0.5 m in accordance with SNiP 2.02.03.

A pile foot reamer is used in competent non-water-saturated soil to increase the bearing capacity by means of increasing the bearing area.

The common disadvantage of the tools shown in Figure 2.1 is the process cyclicity of the soil removal, during which the drilling tool round trips take the most of the working cycle. The use of telescopic drilling rods allows increasing the efficiency due to the reduction in the drilling tool round trip time.

The continuous auger drilling features the high efficiency, with soil being continuously removed to the surface along the blades welded to the hollow core of the auger.

Upon completion of the borehole drilling, a reinforcing cage is introduced and a tremie pipe is lowered into the borehole. Concrete is placed by the vertically moved pipe (VMP). The VMP arrangement includes a hopper (funnel) to load concrete mix; sections of tremie pipes; a lifting beam to support pipes during the assembly (disassembly) process; a lifting element for the assembly (disassembly) of the tremie pipe sections. The pipe diameter is selected depending on the coarse aggregate fraction of concrete mix.

If a pile cuts through soft water-saturated soils, a borehole is drilled using a casing pipe that prevents the borehole walls from collapsing.

### 2.1.1. Piles in boreholes drilled without consolidation of borehole walls

**Shell pile made with the use of multisectional vibration mandrel.** The technology of constructing shell piles in soil was developed by the Kiev Branch of Scientific and Research Hydrotechnical Construction Institute with the participation of the Gersevanov Scientific and Research Institute for Foundations and Substructures. In accordance with the technology, shell piles are made of low-slump concrete mix using the multisectional vibration mandrel.

The vibration mandrel consists of tube sections connected to each other. Circular-vibration generators are installed inside each section. When switched on, each of them acts primarily on the section where it is installed and compacts concrete mix surrounding the section in question.

Shell piles are constructed in stable soil that requires no consolidation of borehole walls. The works are performed in the following order (Figure 2.2). A borehole of required dimensions is driven using a drilling rig. A hopper is installed to receive concrete mix and to center a vibration mandrel. A reinforcing cage is lowered into the borehole. The reinforcing cage is fitted with guide brackets that center the cage relative to the borehole cross section. Concrete is poured in two stages. At first, a lower part of the pile is concreted to a height of 3 to 4 m. The vibration mandrel is lowered on concrete mix, the vibrators are switched on in the lower section and concrete mix is vibro-compacted with the concurrent penetration of the vibration mandrel. When a distance from the lower end of the vibration mandrel submerged into concrete mix is 0.7 of the pile foot diameter (but no shorter than 0.5 m) to the pile foot, the position of the vibration mandrel is fixed by means of the retractable rods of the hopper. Then the upper part of the borehole is filled with concrete mix and vibro-compacted. The vibration mandrel is extracted from the borehole with the running vibrators. The pile hollow is filled with ballast (soil, sand).

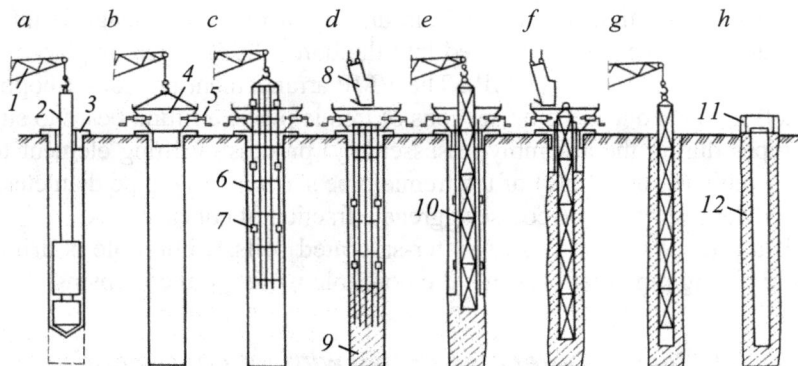

Figure 2.2. Process flow diagram of constructing shell piles
in soil by means of vibration mandrel:
*a* – drilling of boreholes; *b* – installation of hopper; *c* – driving of reinforcing cage;
*d* – concreting of lower part of pile; *e* – lowering of vibration mandrel and
vibro-compacting of concrete mix; *f* – vibro-placing of concrete mix in upper part
of borehole; *g* – extracting of vibration mandrel; *h* – finished pile with formed pile
head; *1* – lifting crane; *2* – drilling string; *3* – conductor; *4* – hopper; *5* – retractable
rod of hopper; *6* – reinforcing cage; *7* – centering bracket of reinforcing cage;
*8* – bucket; *9* – concrete mix; *10* – vibration mandrel; *11* – pile head; *12* – pile

**Vibro-stamped piles** are made in cohesive non-water-saturated soil
that requires no consolidation of borehole walls. Generally, the length of
vibro-stamped piles is not longer than 10 m and the diameter is not greater than 0.5 m.

The technology of constructing such piles includes the following
(Figure 2.3). A borehole of the required overall dimensions is drilled using a drilling rig. At the borehole mouth, a heavy steel conductor is installed that has a hollow guide pipe, the diameter of which is 10 mm
greater than the diameter of the vibro-stamping tool. The conductor
weight is selected so that its pressure on soil is at least 4 to 5 kg/cm$^2$. It is
required to keep concrete mix in the borehole.

The borehole is filled with concrete mix and compacted with the vibro-stamping tool, which is a pipe, whose lower end is closed by a steel
cone. The upper end of the vibration rod is secured to a powerful vibrator.
Due to the fact that the length of the vibration rod exceeds the borehole
depth by 0.8 m, soil under the pile foot is additionally compacted when
the pile is being concreted.

The vibro-stamping tool is sunk into concrete mix with the running
vibrator suspended on the boom of the traveling crane. As a result, an
enlarged base is produced and surrounding soil is additionally compacted.
The pile foot diameter is 30...40 % greater than the pile diameter.

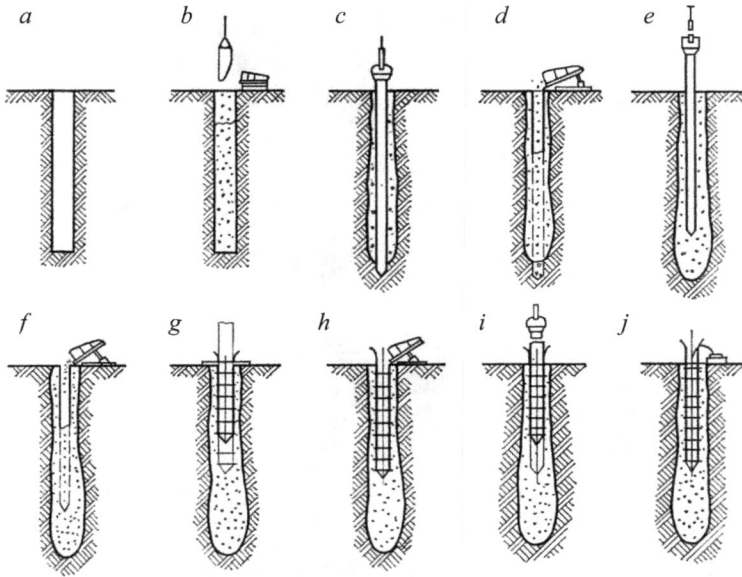

Figure 2.3. Process flow diagram of constructing vibro-stamped piles:
*a* – drilling; *b* – loading of concrete mix from bucket; *c* – driving of vibro-stamping tool into concrete mix; *d* – filling of borehole with concrete mix; *e* – repeated vibro-stamping; *f* – filling of borehole with concrete mix after repeated vibro-stamping; *g* – driving of reinforcing cage into concrete mix by means of vibro-stamping tool; *h* – filling of hollow inside reinforcing cage with concrete mix; *i* – last cycle of vibro-stamping; *j* – final compacting of concrete mix

The concreting operation with the use of the vibro-stamping tool is repeated several times. After the vibration rod has been extracted, a tubular hollow is formed in concrete. A steel cage is inserted in this hollow to provide an interface with a pile cap, and then the hollow is concreted together with the pile cap.

**Piles constructed by means of continuous flight auger.** The abbreviated name of such piles is CFA-piles (continuous flight auger piles) or SOB-piles (from German 'schnecken-ortbeton-pfahl'). The boreholes for these piles are drilled by means of a working attachment – continuous flight auger. Soil is removed from the borehole and delivered to the surface by means of a screw blade welded on the entire length of the auger core (Figure 2.4).

For the pile construction in the confined areas and in case of underpinning, small-scale drilling rigs and short augers are used that form a continuous auger when connected. The small-scale drilling rigs have insignificant impact on bottom soil, which allows constructing piles in the vicinity of existing buildings.

Figure 2.4. Drilling rig UBG-SG Berkut for pile construction using continuous flight auger (*a*) and continuous flight auger (*b*)

*The technology of constructing the piles with the use of continuous flight auger* includes the following:

1. Installation of the drilling rig at the planned location of the pile construction (Figure 2.5, *a*).

2. Lowering of the auger to the design level (Figure 2.5, *b*).

3. Step-by-step extraction of the auger from the borehole and the concurrent filling of the borehole with concrete mix supplied by the concrete pump through the hollow auger (Figure 2.5, *c*).

4. Moving of the rig to the next drilling location (Figure 1.5, *d*).

5. Lowering of the reinforcing cage by means of the vibrator into the borehole filled with concrete mix (Figure 2.5, *e*).

6. Forming of the pile head for the connection to the pile cap (Figure 2.5, *f*).

Depending on its design, two kinds of behavior are possible for the continuous flight auger when it is introduced: partial compaction of soil or certain softening of the 'pile-soil' contact area due to the disturbance in the natural soil structure in drilling operations. For example, the greater is the ratio of the core diameter to the auger blade diameter, the greater is the possibility of soil softening.

The technology is a well-established method for soils, whose layers considerably differs from each other in terms of strength. This technology is especially efficient in the penetration of the thick layers of sand and semi-hard and tough loam when it is not possible to use driven, jacked and cast-in-situ piles.

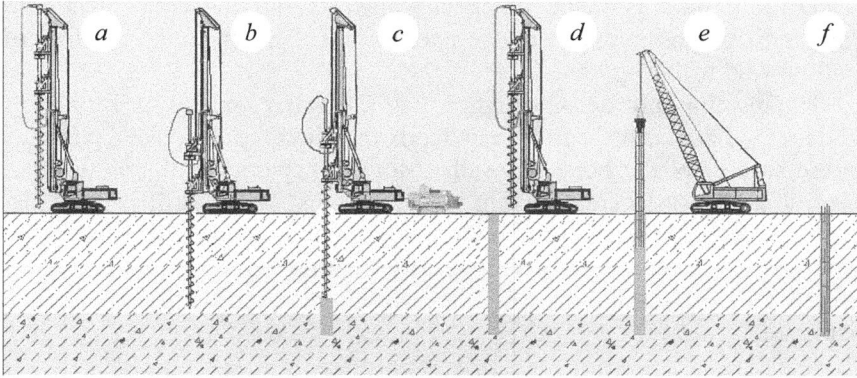

Figure 2.5. Process flow diagram of constructing piles
with use of continuous flight auger

If, in the application of this technology with the continuous construction of a pile field, the thixotropic decompaction of water-saturated clay soil is underestimated for the pile-surrounding soil body, it results in the considerable excess consumption of concrete mix (twice as much and higher). As a rule, the higher consumption of concrete mix occurs when, in the site bed, there are significantly thick layers of flow and very soft loam and sandy clay with low strength and deformation properties [3].

The advantages of this technology are the high efficiency and quality of filling the borehole with concrete mix due to the pressure feed.

The technical features of the Bauer and SoilMec drilling rigs and the range of continuous flight augers are given in Appendices 4 и 5.

## 2.1.2. Piles in boreholes drilled with the use of drilling mud

The method of drilling boreholes with the use of drilling mud was first applied in the penetration of vertical shafts in the beginning of the 1930s.

In the former USSR, the drilling method using drilling mud for bored piles employed in construction was developed in the Scientific and Research Institute for Foundations and Substructures at the S. A. Ter-Galustov's suggestion in 1953 [20].

In unstable and water-saturated soil, the internal pressure higher than the external pressure is built to prevent the walls from collapsing in the boreholes. To this effect, the borehole is continuously refilled with water so that the water level is several meters higher than the groundwater table or the boreholes are filled with drilling mud.

Drilling mud is made of special-purpose fine-dispersed bentonite clay, and it is possible also to use native clays. The drilling mud density shall be 1.05 to 1.25 $g/cm^3$.

Having the bulk density higher than the water density, drilling mud produces the excess pressure at any depth, and therefore holds soil particles on the surface of the borehole walls. Moreover, particles of drilling mud consolidate the walls creating thin but stable crust. When drilling mud circulates, loosened material is brought from the borehole to the surface.

The pile-concreting is performed by the method of vertically moved pipe (VMP). As the borehole is filled with concrete mix, drilling mud is displaced.

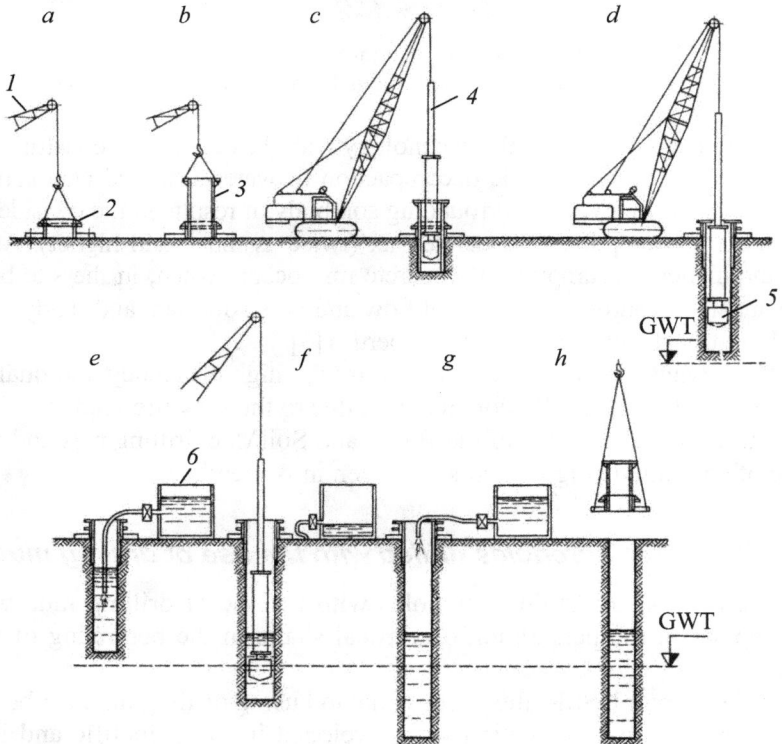

Figure 2.6. Process sequence of drilling boreholes with use of drilling mud in unstable soil: *1* – boom of self-propelled crane; *2* – conductor; *3* – nipple; *4* – drilling rig; *5* – drilling tool; *6* – container for drilling mud; GWT = groundwater table

*The process of drilling boreholes with drilling mud for the pile construction* includes the following operations:

1. Installation of drill conductor (Figure 2.6, *a*).
2. Installation of casing nipple (Figure 2.6, *b*).
3. Setting of casing nipple into borehole (Figure 2.6, *c*).
4. Beginning of borehole drilling within layers of competent soil (Figure 2.6, *d*).
5. Filling of borehole with drilling mud (Figure 2.6, *e*).
6. Drilling operation using drilling mud (Figure 2.6, *f*).
7. Refilling of brehole with drilling mud (Figure 2.6, *g*).
8. Removal of drill conductor and casing nipple (Figure 2.6, *h*).

To perform the drilling operations, a drilling mud container and waste slurry sump shall be provided on the site.

Drill cuttings are removed with drilling mud by means of sludge pumps. Drilling mud carries away drill cuttings and, at the same time, cools down the drilling tool. Upon completion of drilling, drilling mud is left in the borehole to prevent the walls from collapsing and creeping. The borehole depth is controlled by means of a special-purpose plumbbline with a weight of 2 to 3 kg; in addition, controlled are the borehole diameter and the availability of the casing nipple in the borehole mouth. An authorization to install a reinforcing cage is given based on the survey results.

*The process of pile concreting* under drilling mud includes the following operations:

1. Installation of casing nipple (Figure 2.7, *a*).
2. Installation of reinforcing cage (Figure 2.7, *b*).
3. Installation of sleeve for hopper (Figure 2.7, *c*).
4. Lowering of tremie pipe (Figure 2.7, *d*).
5. Fastening of hopper to pipe (Figure 2.7, *e*).
6. First stage of borehole concreting (Figure 2.7, *f*).
7. Continuing of concreting with lower sections of tremie pipe removed (Figure 2.7, *g*).
8. Filling of borehole with concrete while hopper with tremie pipe is being lifted (Figure 2.7, *h*).
9. Removal of sleeve (Figure 2.7, *i*).
10. Removal of casing nipple (Figure 2.7, *j*).
11. Installation of ready-made formwork for pile head (Figure 2.7, *k*).
12. Concreting of pile head (Figure 2.7, *l*).
13. Removal of formwork from pile head (Figure 2.7, *m*).
14. Covering of pile head (Figure 2.7, *n*).

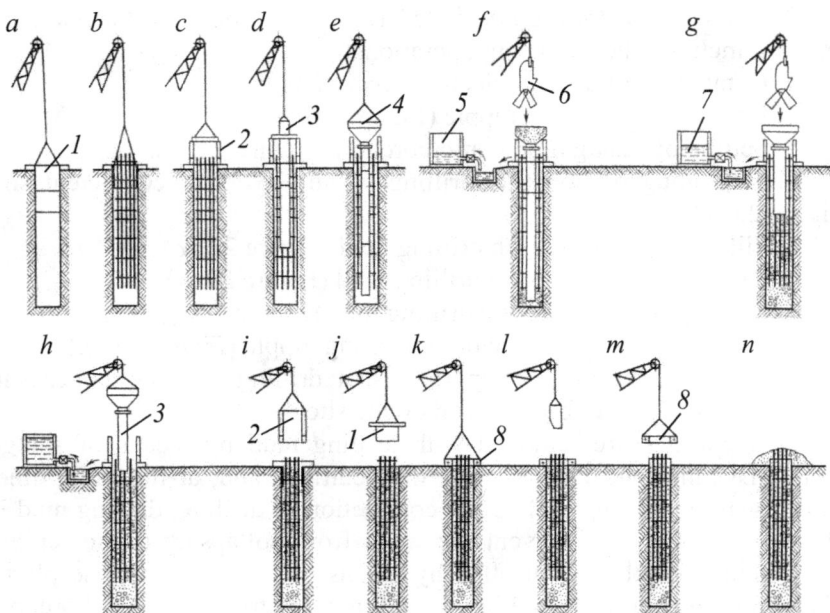

Figure 2.7. Process flow diagram of concreting piles in borehole with drilling mud:
1 – casing nipple; 2 – sleeve for hopper; 3 – casing pipe; 4 – hopper; 5 – sump;
6 – bucket; 7 – slurry; 8 – pile head formwork

The reinforcing cage installed in the borehole is held suspended on the casing nipple by means of steel bars.

A pipe to feed concrete mix is installed inside the borehole to the entire borehole depth. The telescopic tremie pipe for dry boreholes shall not be used when concreting piles in the boreholes under drilling mud. Sectional pipes shall have airtight joints.

A hopper is fastened to the tremie pipe, and the hopper capacity shall not be smaller than the capacity of the tremie pipe. As the borehole is filled with concrete mix, the pipe is moved upward by a crane, thus providing the continuity of the concreting process. In all cases, a level of concrete mix in the pipe shall not be higher than a level of drilling mud in the borehole, so the pipe is lowered into concrete mix at least by 1 m.

In the process of concreting, drilling mud is displaced by concrete mix through the casing annulus to the borehole mouth from where it discharged via chutes to the sump for cleaning and reuse. When drilling mud is saturated with cement, it is drained to a dump pit. At the upward flow velocity of drilling mud of 0.6 to 0.7 m/s, drill cuttings are brought out to the surface.

The concreting by the VMP method is carried out until concrete mix appears on the surface, and then the contaminated layer of mix is removed. Upon removal of the casing nipple, the formwork is installed and the pile head is concreted.

When performing the works, no breaks in concreting are allowed with the duration exceeding the beginning time of concrete setting, and the works shall not be resumed until the hardened surface of concrete is properly prepared and jelled drilling mud is removed.

Breaks with duration shorter than the beginning time of concrete setting do not interfere with continuing the works, but before the next concreting operation, it is required to provide the circulation of drilling mud, for example, by way of direct flushing via a pipe lowered down to the bottom. The circulation is required for suspending drilling mud particles that precipitate on the bottom, reduce the vertical and horizontal dimensions of the enlargement and create an undesirable intermediate layer between the pile enlarged foot and soil where it shall be placed.

To facilitate the pile concreting, the boreholes should be drilled using the flushing with lighter and less viscous drilling mud, which is acceptable for soil under consideration, and in individual cases, the flushing should be done with water.

### 2.1.3. Piles in boreholes drilled with casing pipe

When constructing piles in water-saturated soil, the most reliable protection against the collapse of the borehole walls is the use of casing pipes. Such piles were first designed by Kiev Engineer A. E. Strauss in 1899. The boreholes for piles were drilled manually using a gradually added-on steel casing pipe 30 to 40 cm in diameter. Spoon bits (coiled pipes) or drilling bits (chisels) were used depending on the soil properties. To raise and lower a drilling tool over the pile-construction location, a tripod was installed on a pulley (Figure 2.8). The other end of a rope was reeled on the winch drum.

The casing pipe was lowered as soil was excavated. In sand and running soil, boreholes were drilled using a chisel fitted with a socket. The purpose of the socket is to increase the chisel weight and the force of its impact when the chisel falls down.

In the construction of the Strauss piles, the boreholes were drilled with the insignificant use of machinery, due to which the pile length was limited and did not exceed 10 to 12 m.

Prior to concreting, the borehole bottom was cleaned with a spoon bit. Concrete mix was delivered in bottom dump buckets. After the next batch has been loaded, concrete mix was thoroughly compacted with the concurrent extraction of the pipe.

To maintain the integrity of the pile shaft, the casing pipe was raised to a height of ¾ of concrete layer. Under action of compacting, the pile shaft enlarged and took an irregular form. This being the case, the pile thickness in the cross section was inversely proportional to the soil density. The pile diameter enlarged by 30 to 50 % of the casing pipe diameter, and, accordingly, the consumption of concrete mix increased sometimes reaching the

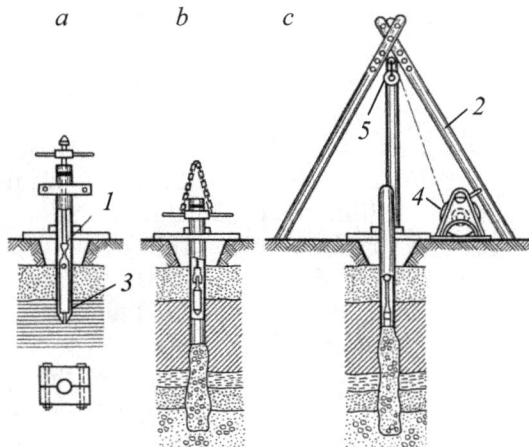

Figure 2.8. Process flow diagram of constructing Strauss pile: *a* – drilling of borehole; *b* – feeding of concrete mix to pipe; *c* – compacting of concrete mix; *1* – chisel; *2* – tripod; *3* – casing pipe; *4* – drum; *5* – pulley

three-fold volume. Dry-concrete mix was used to make piles in dry soil, while flow concrete was used in the presence of groundwater.

It is difficult to compact concrete mix in water as the movement of tampers makes water turbid, concrete mix disintegrates, and dirt seams occur in a pile. For this reason, flow concrete started to be used for the pile construction in water-saturated soil, which allowed eliminating the concrete disintegration.

In the presence of groundwater, casing pipes shall be filled with concrete mix to the entire height in a single step. Concrete mix comes out of the pipe while it is being extracted and, under the pressure of overlying column, compacts soil by filling up the borehole.

The Strauss piles were successfully used at a number of major projects in the years of the first 'five-year plans'.

One of the main advantages of the Strauss piles is the possibility to lower the casing pipe without impacts and shakes. It is very important in the applications where the pile foundations are constructed in the vicinity of the buildings and structures that are sensitive to dynamic loads, as well as inside buildings.

Among the disadvantages of Strauss piles are the low efficiency of works due to the manual drilling operations and the difficult control of the shaft integrity in the presence of groundwater.

Currently, even though Strauss piles are not used in their original form, a family of improved piles was designed on their basis. The borehole-drilling equipment, the methods of casing pipe penetration and extraction, concrete mix feed and compaction have been improving with time.

**Piles in boreholes drilled with casing pipe by Kelly method.** The Kelly method implies the soil excavation using a drilling tool (short auger, drill bucket or rotary core bit) attached to the tip of a telescopic bar (Kelly bar).

A casing pipe is lowered by way of rotating through a clamp fastened to the pipe and jacking by means of a hydraulic jack. As the casing pipe is lowered, soil is removed from it and the next section is added on it. Soil is excavated with a short auger fastened to the tip of the telescopic bar (Figure 2.9). Concrete mix is delivered to a tremie pipe from the mixer truck chute. To prevent water from entering the borehole, the joints of the sections are sealed with bitumen-impregnated cardboard. Generally, the thickness of the pipe wall is 4.0 mm. To a great extent, the penetration rate depends on a type of drilling equipment.

Figure 2.9. Drilling of borehole with casing pipe:
*1* – casing table; *2* – casing pipe; *3* – short auger; *4* – telescopic Kelly bar;
*5* – rotary disc; *6* – mast; *7* – Bauer drilling rig

*The technology of constructing the piles with the use of the casing pipe* includes the following:

1. Positioning of the drilling rig at the pile construction location (Figure 2.10, *a*).

2. Lowering of the casing pipe section by section to the required depth; removal of soil from the casing pipe (Figure 2.10, *b*).

3. Extraction of the auger from the casing pipe, removal of sludge from the borehole bottom by means of the drilling bucket, lowering of the reinforcing cage (Figure 2.10, *c*).

4. Concreting of pile (Figure 2.10, *d*) by the method of vertically moved pipe (VMP). The tremie pipe is assembled of sections and lowered into the casing pipe to the entire depth.

5. Extraction of the casing pipe (Figure 2.10, *e*).

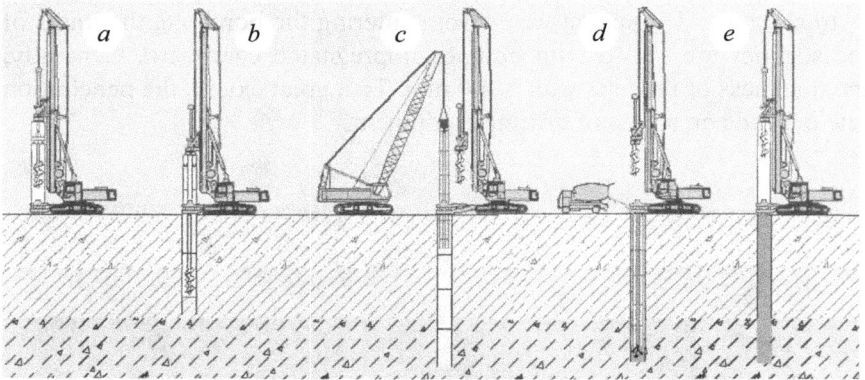

Figure 2.10. Process flow diagram of constructing pile in borehole drilled with casing pipe by means of short auger and telescopic bar

The technology has a number of advantages:
- as there no dynamic loads and vibration impact on soil, piles can be constructed in the vicinity of existing buildings and structures;
- state-of-the-art equipment of the drilling rig makes it possible to control the drilling process, to drill out and remove boulders;
- filling of the borehole through the tremie pipe eliminates the necking;
- in the process of drilling, the direct control is effected for the compliance of the actual soil properties with those specified in the design;
- possibility to make an enlargement provides the most complete use of the bearing capacity of the soil base of the pile.

The technology disadvantage is the low efficiency; therefore the design solutions shall provide the full use of the soil-specific bearing capacity of piles. Besides, it is require to make a soil plug of considerable length or to build up the excess pressure in the borehole using water or drilling mud in order to prevent water-saturated soil from heaving into the borehole when excavating through water-saturated soil.

To excavate boreholes by the Kelly method, there are commonly used the Bauer and SoilMec rotary drilling rigs, whose technical features are given in Appendices 4 and 5.

**Piles in boreholes drilled with continuous flight auger with use of casing pipe** This technology is also called 'double rotary', FOW system (from 'front-of-wall system') or VDW system (from German vor-der-wand system).

When constructing piles by the double-rotary technology (Figure 2.11), the boreholes for the piles are drilled using a traveling casing pipe, with the right-hand rotation of the continuous flight auger inside the casing pipe and the left-hand rotation of the pipe being performed at the same time. Figure 2.12 shows a drilling rig with a working tool to make the above piles.

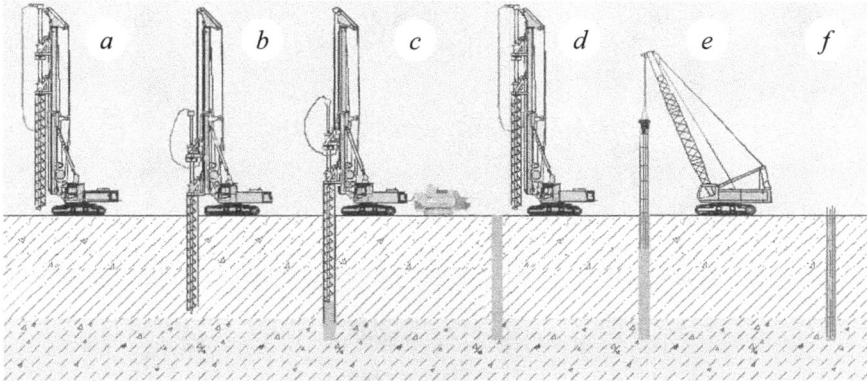

Figure 2.11. Flow diagram of constructing piles using double-rotary technology

*The technology of constructing the piles by the double-rotary method* is as follows:

1. Positioning of the drilling rig at the drilling location (Figure 2.11, *a*).

2. Drilling of boreholes (see Figure 2.11, *b*) to the required depth with the concurrent penetration of the continuous flight auger (right-hand rotation) and the casing pipe (left-hand rotation).

3. Concreting of the pile through the auger core with the concurrent raising of the auger and casing pipe (see Figure 2.11, *c*).

4. Removal of soil from the casing pipe at the left-hand rotation of the auger (see Figure 2.11, *d*).

5. Lowering of the reinforcing cage into the borehole filled with concrete mix by means of a vibrator suspended on the crane boom (see Figure 2.11, *e*).

6. Forming of the pile head for the connection to the pile cap (see Figure 2.11, *f*).

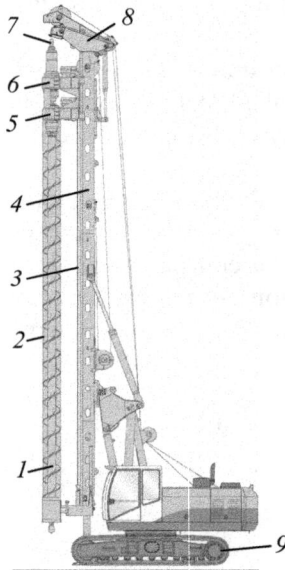

Figure 2.12. Drilling rig with equipment to drill boreholes
with continuous flight auger with use of casing pipe:
*1* – continuous flight auger; *2* – casing pipe; *3* – winch to feed drilling tool downward; *4* – standard mast of drilling rig; *5* – auger head; *6* – rotary head for casing pipe; *7* – nipple to feed concrete mix to auger core; *8* – jib;
*9* – undercarriage

Below listed are the advantages of the technology:
- technology is applicable to dispersed sol of all kinds (noncohesive compact soil, silt, hard clay);
- no noise and significant vibration impact (that allows constructing piles in the vicinity of exiating buildings);
- high efficiency due to the soil excavation by means of continuous flight auger;
- high quality of filling a borehole with concrete mix due to the feed of concrete mix under pressure;
- possibility to control the drilling parameters with on-board computer.

Appendix 4 (Table A.4.3) shows the parameters of the borehole excavation using the double-rotary technology by means Bauer drilling equipment and rigs.

### 2.1.4. Piles with enlargements

**Methods of making enlargements.** Different methods are used to make an enlargement in the cross section of a pile. Among them, there is drilling-out, pressing or reaming of soil, treatment of concrete mix with electric discharges. A pile foot can be enlarged as a result of the charge explosion or the tamping of dry-concrete mix at the borehole bottom.

The drilling-out implies an enlargement formed in the borehole by means of cutting soil with a special-purpose tool – reamer. A reamer is fitted with blades that cut soil when the reamer rotates. Cut soil gets into a soil receiver of the reamer and withdrawn from the borehole to the surface.

Figure 2.13, *a* shows a reaming bit, which is a cylinder with retractable blades. The cylinder diameter corresponds to the borehole diameter for a pile. Two cutting blades are mounted inside the cylinder and connected to each other by an arm system. After the borehole for the pile has been finished, the reaming bit attached to the tip of a drilling rod is lowered into the borehole. During the penetration, the cutting blades are inside the cylinder. Once the cylinder supports on the borehole bottom, a pressing force is applied to the drilling rod. The drilling rod applies pressure to the arm system and the cutting blades move apart going beyond

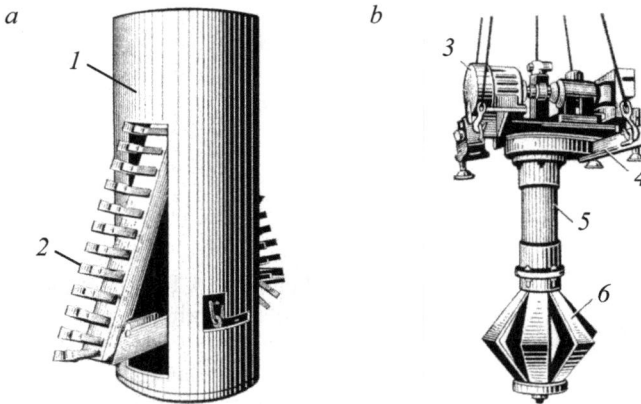

Figure 2.13. Attachments to enlarge boreholes by way of drilling-out (*a*) and static pressing (*b*) of soil:
*1* – cylinder of reaming bit; *2* – blade; *3* – electric motor of unit for biconic-shaped borehole reaming; *4* – supporting frame; *5* – rod; *6* – stamping plates

the cylinder. Then a torque is applied to the drilling rod, and the drilling tool starts rotating under the action of the torque. During the rotation, soil is cut and gets into the cylinder. After it has been filled up, the cylinder is withdrawn to the surface and unloaded.

Pressed enlargement are made by way of pressing in soil into the borehole walls with the stamping plates of the reaming bit (Figure 2.13, *b*). There are also used the shells, which expand when cement grout is injected into them.

When making an enlargement by reaming, a load on soil is transferred by rollers of special-purpose tools. Those tools are used as attachments of the rotary drilling rigs. Figure 2.14 shows a URS-1M rolling reamer for boreholes designed by the Kazakhstan Designing and Technological Institute for Foundation Construction. It allows making enlargements 1200 mm in diameter at any depth of a borehole 600 mm in diameter. The URS-1M reamer includes a Kellybar that passes through a hole in the drilling column and a hinged system with two pairs of rollers: rolling-out and rolling-down. The hinged system has also supportingrollers that move along the guide retainer, through which the rod passes. To make the rolling-down rollers move, pin-jointed arms are used that open when the drilling column moves downward under the action of axial force. The retainer has two pairs of trunnions that fix it to the rod when the reamer is running and fix the rod in the drilling column when the reamer is being extracted from the borehole. The axle of the hinged system together with the hub is capable to travel longitudinally relative to the retainer and rod. The hubis connected to the drilling column by means of a lock.

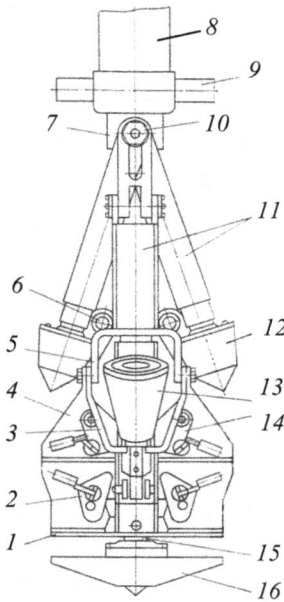

Figure 2.14. Design of tool for borehole enlarging by way of reaming:
1 – disc; 2 – rod trunnions; 3 – hinge arm; 4 – retainer; 5 – hinge rockers; 6 – track roller; 7 – hub; 8 – drilling column; 9 – lock; 10 – axle; 11 – hinged system; 12 – rolling-out rollers; 13 – rolling-down rollers; 14 – retainer trunnions; 15 – rod; 16 – centering support

**Piles with conic and bi-conic enlargements.** Enlargements of piles can be achieved by way of drilling out or pressing in soil. For the drilling-out, there are used the reaming bits (see Figure 2.13, *a*) or pantograph-type reamers. Soil is pressed into the borehole walls by means of special-purpose tools fitted with stamping plates (see Figure 2.13, *b*).

A pantograph reamers (Figure 2.15) includes a cutting mechanism and a soil-collector. The cutting mechanism is a system, which is fitted with the pin-connected blades and connected to the tip of the drilling rod. After the reamer has been lowered into the borehole bottom, a pressing force is applied to the drilling rod,

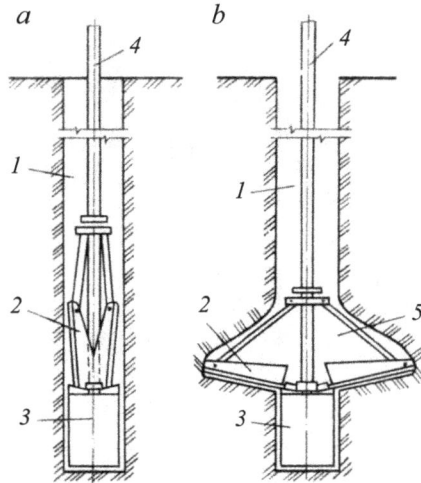

Figure 2.15. Drilling-out of soil with pantograph-type reamer: *a* – position of reamer in borehole prior to drilling-out of soil; *b* – the same after drilling-out of soil; *1* – borehole; *2* – cutting blades of reamer; *3* – soil collector; *4* – rod; *5* – enlargement

and the cutting blades fitted to the tip expand under the action of this pressing force. Then the soil cutting starts by rotating the rod. A cylindrical bucket is mounted under the cutting mechanism to collect cut soil. The bucket bottom is provided with cleaning knives.

A disadvantage of the reaming bits and pantograph reamers is the cyclicity of the process, during which the drilling tool round trips take up to 30 % of the time.

**Piles with camouflet enlarged footing.** In the explosion method, an enlargement in the borehole hollow is formed as a result of confined explosion. The hollow dimensions depend on the soil properties, the amount and type of the explosive.

The enlargement of pile bases by the confined explosion was first performed by Wilhelmi in 1901. The boreholes for Wilhelmi piles were made using casing pipe. After the pipe has been extracted from soil, a blasting charge was lowered onto the borehole bottom and the borehole was filled with concrete mix. The casing pipe performed a stemming function. The explosion resulted in an enlarged hollow. Then the pile concreteing was finished and the casing pipe was extracted.

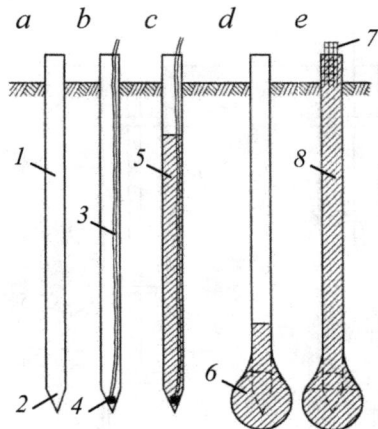

Figure 2.16. A.A. Luga's process flow diagram for constructing piles with camouflet enlarged footing: *1* – hollow shell; *2* – tapered tip; *3* – explosion system; *4* – blasting charge; *5* – concrete mix; *6* – camouflet enlarged footing; *7* – reinforcing cage; *8* – pile

In 1941, A. A. Luga suggested another method of constructing piles with camouflet enlarged footing in order to eliminate the time-consuming operation of soil removal from the casing pipe. In accordance with the suggested technology, a closed-end steel shell was lowered into soil, and a charge was placed in the center of the tapered tip of the shell (Figure 2.16). When calculating the charge, A. A. Luga considered additionally the necessity for the explosion of the shell tip. The charge was fired after the shell has been filled up with concrete mix. Then a reinforcing cage was lowered into the borehole and the remaining part of the borehole was filled up with concrete mix. In our country, the above piles were used as deep-foundation piles in the construction of bridge pillar in the middle of the XXth century.

With time, the technology had been slightly changed depending on the soil capability to retain the borehole walls, the foundation depth, the pile materials and design. However, the principal sequence of the operations has remained the same.

**Bored down-take piles with camouflet enlarged footing.** To improve the reliability of a bored pile, a prefabricated pile is lowered, after explosion, into a borehole partially filled with concrete mix. Figure 2.17 shows a sequence of the operations related to the construction of the above piles.

It is recommended to make camouflet enlarged footing when the lower part of a pile is in competent cohesive soil. In noncohesive and low-cohesive soil, boreholes shall be cased with standard casing pipes. The casing pipes should be lowered not to the entire borehole depth but 0.8 to 1.2 m higher than the bottom to prevent the pipe end from deforming at the time of explosion. No camouflet enlarged footing can be provided in water-saturated silty sand, flow and very soft clay soil, as well as in rudaceous and rocky soil.

The advantage of this technology is the higher bearing capacity due to camouflet enlarged footing that increases the area of pile bearing on soil.

The technology has a number of disadvantages that considerably reduce the application of piles with camouflet enlarged footing. For example, these piles cannot be used in confined areas of compact development and in the vicinity of explosion-hazardous production facilities. It is required to provide a storage area to store explosives on a construction site. This technology does not allow effecting the reliable quality control of the con-

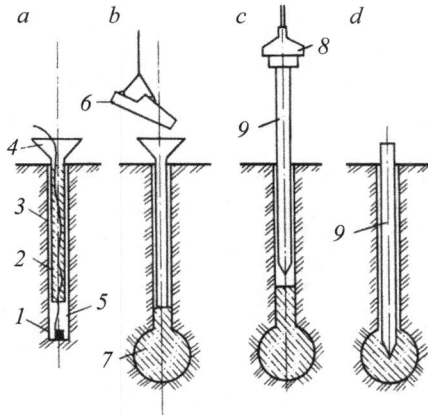

Figure 2.17. Process flow diagram for constructing bored down-take piles with camouflet enlarged footing:
*1* – charge; *2* – wire to blasting machine; *3* – casing pipe; *4* – funnel; *5* – concrete mix; *6* – bucket; *7* – camouflet enlarged footing; *8* – vibratory pile driver; *9* – reinforced concrete pile

struction of camouflet enlarged footing. After explosion, the upper crest of enlargement might collapse and concrete mix might disintegrate. To construct piles with camouflet enlarged footing, it is required to obtain a permit for blasting operations that shall be performed by a specialized company.

**Construction of piles using pulse discharge technology.** G. M. Lomize suggested the use of the electric-charge technology in the geotechnical engineering to compact water-saturated sand, sandy clay and loess-like loam. P. L. Ivanov made a substantial contribution to solve a problem of compacting the water-saturated sand with momentary dynamic action. The principles of the technology to construct piles using the electrohydraulic effect were developed in the Leningrad Institute of Civil Engineering in 1978–1981.

The essence of the pulse discharge technology lies in the following: a borehole filled with fine-aggregate concrete or cement grout is treated with a series of high voltage electric discharges. In doing so, the electrohydraulic effect occurs that results in forming a pile shaft or anchor root and in cementing and compacting surrounding soil. As a result of the treatment with the design series of discharges, the initial borehole diameter (130 to 300 m) can be increased more than 2 times depending on the energy supplied to the bo-

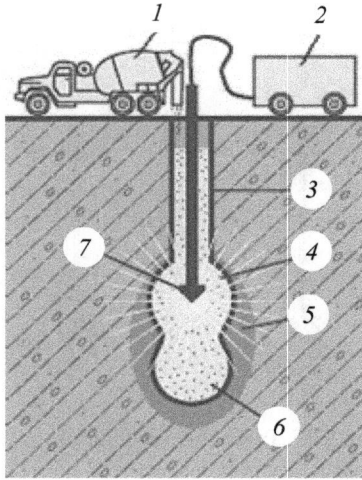

Figure 2.18. Forming of pile enlargement using pulse-discharge technology: *1* – mixer truck; *2* – surge-current generator; *3* – borehole before treatment; *4* – area of soil cementation; *5* – area of soil compacting; *6* – enlargement at pile base; *7* – emitter

rehole and the hydrogeological site conditions. Surrounding soil is compacted and the porosity is reduced in the shock pulse impact area.

To treat concrete mix or cement grout with electric discharges, a surge-current generator (SCG) is used that switches on a transformer, rectifier, energy storage unit, switch gear and control unit. The generator is connected to an energy emitter installed in the borehole filled with concrete mix (Figure 2.18). Discharge is generated as follows. The voltage of DC electric energy is increased from 220...380 V to 10 kV. High-voltage DC electric energy is accumulated in the energy storage unit, which is a capacitor bank. This energy is directed to the emitter submerged into concrete mix or cement-and-sand grout. When electric energy is supplied to the emitter electrodes, high-density energy is generated in the interelectrode gap and a breakdown occurs with creating the plasma channel of the discharge where the temperature increases up to $10^4...10^5$ °C and the pressure increases up to $10^8...10^{13}$ Pa in $10^{-4}...10^{-5}$ s. At that, compression waves form and spread in the surrounding environment. At this stage, the stored electric energy transforms into the electrodynamic disturbance energy, which results in the expansion of the discharge channel to a gas-vapor cavity. When the pressure in the cavity becomes lower than the hydrostatic pressure of concrete mix, the cavity starts collapsing. After the discharge, the soil compaction degree is estimated by a settlement of concrete mix relative to the borehole mouth.

To construct piles and to compact soil, the electric discharge energy of 20 to 60 kJ per pulse is used at the discharge frequency of 3 to 20 pulse per minute. The electric discharge energy of 5 to 15 kJ is specified when cement-grouting the foundation-base contact area.

When using the electric discharge energy up to 60 kJ, the dynamic action occurring beyond the treatment area is insignificant and have no adverse effect on the structures to be strengthened and on the adjacent buildings. The pulse-discharge technology is environmentally safe; it allows constructing piles and anchors of various shapes and with enlargements on one or multiple levels.

The piles made by this technology are called, in abbreviated form, PDT (Pulse-Discharge Technology) piles. In 1993, the Gersevanov NIIOSP developed an instruction for the use of the pulse-discharge technology in the pile construction. In 1997, the Recommendations for Application of Bored Injected Piles were issued that specify the construction technology and the design procedure for the above piles.

*The PDT pile technology* includes the following operations:

1. Drilling of a borehole.

2. Filling of the borehole with concrete mix or cement grout.

3. Treatment of concrete mix or cement grout using the electric-discharge technology on the required depth.

4. Lowering of a reinforcing cage into the borehole.

The pulse-discharge technology has the following advantages:

• use of light-weight small-scale machines allows performing the works in a basement (at least 2.4 m high), basement storey or first floor without inconvenience for the residents of overlying floors and adjacent buildings;

• boreholes can be drilled without casing pipes in unstable soil with the boreholes walls collapsing.

The bearing capacity of the PDT piles 1.5 to 2.5 times higher than the bearing capacity of bored piles made with the use of the conventional technologies.

The high bearing capacity of the PDT piles is due to the enlargement of the borehole shaft; the soil compaction around the shaft and nuder the pile foot; the partial cementation of sandy soil around the shafts.

The soil resistance increases by 1.3 to 2.0 times under the pile foot and by 1.2 to 1.5 times along the side surface.

One type of the electric-discharge technology is the magnetic-pulse treatment of concrete mix that allows considerably increasing the strength and homogeneity of concrete mix, and the quality and reliability of a pile.

The PDT piles are successfully used in the reconstruction of the existing building and structures and in the construction of new ones. The recommended inclination to the vertical line is no more than 20°.

The applications of the PDT piles are the following:

1. Construction of pile foundations when constructing new buildings in confined areas in close vicinity to the existing buildings.

2. Construction of enclosing structures similar to secant pile walls and diaphragm walls.

The use of the PDT piles in the enclosing structures allows achieving, with the minimum soil excavation in drilling, a structure, which is practically as good as a diaphragm wall in terms of rigidity and permeability and, moreover, capable to bear a sufficiently high vertical load. Due to the fact that soil around the piles is compacted and sand is cemented as well, it becomes possible to construct piles at a relatively close distance from each other.

3. Underpinning of existing foundations by way of transferring the entire or partial load from a structure to piles when changes are made in architectural-and-planning and structural solutions for the existing buildings (overstorey, increasing of spans and loads, increasing of basement height).

If required, the pulse-discharge technology can be used for cementing the brickwork and stone masonry of foundations, this being a case, when the electric-discharge energy is specified within a range of 0.3 to 1.5 kJ, and the discharge frequency is specified within a range of 10 to 150 discharges per minute.

When underpinning the existing foundations, the structural solutions are similar to those used for bored injected piles. In new construction, the structural solutions for the PDT piles have no specific features. In new construction and in the underpinning of existing foundations, the distinctive feature of the PDT pile application is the possibility to achieve a high bearing capacity of a pile with the minimum drilling diameter and length. The piles of the drilling diameter of 150 to 250 mm have the bearing capacity not lower than that of the driven piles with the cross section of 300×300 mm and of the same length.

## 2.2. Displacement piles

Displacement piles are concreted in boreholes resulted from the forced displacement of soil. The boreholes can be made by way of percussion, reaming, static or vibration displacement of soil.

Mandrels or tubes with a closed lower end are used for percussion and static or vibration displacement of soil. Depending on the pile-constructing method, the mandrels and tubes are lowered by means of vibratory pile drivers, pile-driving or pile jacking rigs.

Reaming (screw pressing-in) of a borehole is done by way of screwing into the ground a helical tool or a tube, whose lower end is closed with a screw tip to be left in the ground.

The possibility of using one or another borehole-making method for the displacement piles shall be determined at the stage of the engineering and geological survey. Table 2.1 shows the application of different borehole-making methods for the displacement piles depending on soil conditions.

*Table 2.1*

**Methods of borehole-making for displacement piles depending on soil conditions:**

| Borehole-making method | Soil conditions | |
|---|---|---|
| | Clay soils with index of liquidity $I_L$ | Sand |
| Static displacement | $I_L > 0.3$ | – |
| Reaming | $I_L > 0.2$ | Silty soft water-saturated |
| Vibration displacement | $I_L > 0.6$ | Soft water-saturated |
| Percussion | $I_L > 0.3$ | Fine- and medium-grade |

The engineering and geological survey shall identify the presence or absence of soils at the bottom that are capable to be broken in case of the reaming or static displacement or to be restructured in case of the percussion or vibration displacement. The breaking and restructuring are accompanied by the sudden deterioration of the mechanical properties: reduction in strength and increase in compressibility of soils.

If there thixotropic soils are present, it is required to determine their sensibility to the disturbance of the structural bonds, i.e., the degree of the deterioration of the mechanical properties (strength and deformation) of soils and the bearing capacity of piles resulted from the borehole making. There also shall be determined the time of the structural bond restoration and the degree of the soil strength increase.

In case of the occurrence of the thick sand soil layer, it is required to estimate their degree of compaction resulted from the percussion or vibration displacement of borehole.

### 2.2.1. Piles in boreholes made by percussion

*Piles in boreholes made by driving of mandrels*

The **Compressol piles** were developed by French Engineer Dulac in 1900. The construction technology of those piles includes three operations (Figure 2.19). A well for a pile is made by way of dropping a heavy cone-shaped cast-iron tamper from a certain height. Upon reaching the required depth, the borehole bottom is filled with crushed rock or dry-concrete mix and tamped, thus providing an enlargement in the lower section of the borehole shaft. Then the remaining part of the borehole is filled with concrete mix and compacted by a tamper with the flat working surface.

As the Compressol pile construction technology did not involve the use of casing pipes, it was used in cohesive soil that is capable to retain the borehole walls vertical.

## *Piles in boreholes made by driving of reusable pipes, whose lower end is closed with shoe to be lost in ground*

The **Simplex piles** – piles with a driven casing withdrawn out of the ground – were invented by Designer Frank Shuman and first used in USA in 1903. Figure 2.20 shows the sequence of the operations to construct such piles.

*The technology of Simplex displacement pile construction* includes the following:

1. Driving of a thick-walled casing pipe into the ground; the casing pipe has a diameter of 400 mm and its bottom is closed with a cast iron shoe (Figure 2.20, *a*).

2. Batch supply of concrete mix to the casing pipe using a hinged-bottom bucket (Figure 2.20, *b*).

3. Tamping of concrete mix using the tamper with the casing pipe being withdrawn at the same time (Figure 2.20, *c*).

4. Withdrawal of the casing pipe; forming of the pile head (Figure 2.20, *d*).

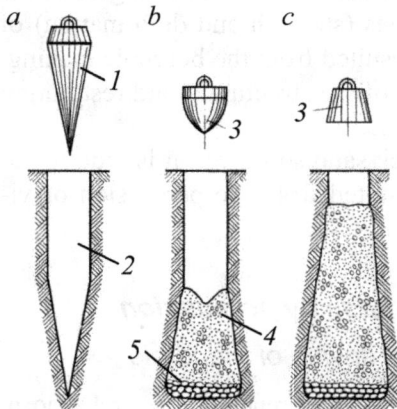

Figure 2.19. Process flow diagram of constructing Compressol pile: *1* – cone for borehole percussion; *2* – borehole; *3* – tamper; *4* – concrete mix; *5* – crushed rock or dry-concrete mix

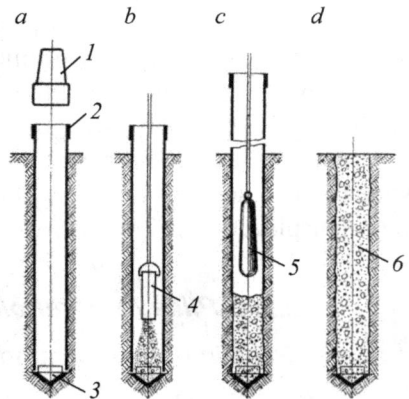

Figure 2.20. Process flow diagram of constructing Simplex piles: *1* – hammer; *2* – casing pipe; *3* – shoe; *4* – bottom dump bucket; *5* – tamper; *6* – pile

The **Harley Abbott piles** differ from the Simplex piles with an enlarged base. In a thick-walled casing pipe 40 cm in diameter, a mandrel is inserted that projects 1.25...1.50 m down beyond the pipe. The enlarged mandrel head is supported by the casing pipe. The impact energy is transferred through the enlarged head to the casing.

The technology of the Harley Abbott pile construction includes the following:

1. Driving of a mandrel into the ground (Figure 2.21, a).

2. Withdrawal of the mandrel; supply to the pipe a batch of concrete mix, pulling back of the pipe (Figure 2.21, b).

3. Forming an enlargement by means of the mandrel hitting concrete mix (Figure 2.21, c).

4. Completion of concreting; withdrawal of the casing pipe; forming of the pile head (Figure 2.21, d).

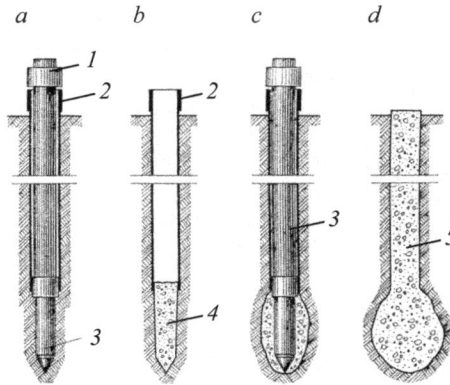

Figure 2.21. Process flow diagram of making Harley Abbott piles:
1 – mandrel head; 2 – casing pipe; 3 – mandrel; 4 – concrete mix; 5 – pile

Being designed in France in 1909, the **Frankignoul piles** soon started to be widely applied (Figure 2.22). A casing pipe consists of concentric rings inserted into each other. A length of the rings is 3 to 6 m with a diameter being 40 to 70 cm. Special-purpose couplings prevent the rings from being disconnected without interfering with the telescopic movement of the rings.

A pipe is driven by a tip with a rod, which serves as guide for the hammer travel.

The technology of Frankignoul pile construction includes the following:

1. Driving of the first ring of the casing pipe (Figure 2.22, a).

2. Driving of the second ring of the casing pipe (Figure 2.22, b).

3. Withdrawal of the hammer together with the tip from the casing pipe and the beginning of concreting (Figure 2.22, c).

4. Tamping of concrete in the pipe, which is slightly lifted upward. Under the tamping pressure, concrete is pressed into the ground with enlarging the pile cross section. The tamper travels along the guide rod, which is gradually being embedded in concrete, while the lower ring of the casing pipe is lifted upward by means of tension bars; the lower hooks are inserted into the holes of the casing pipe (Figure 2.22, d).

5. Withdrawal of the lower ring of the casing pipe and the beginning of concreting the next ring (Figure 2.22, *e*).

6. Completion of concreting; withdrawal of all rings of the casing pipe from the ground; forming of the pile head (Figure 2.22, *f*).

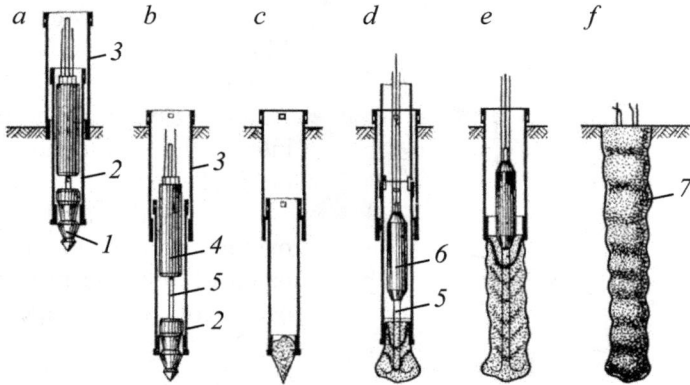

Figure 2.22. Process flow diagram of constructing Frankignoul piles:
*1* – tip; *2* – first ring of casing pipe; *3* – second ring of casing pipe;
*4* – hammer; *5* – guide rod; *6* – tamper; *7* – pile

The **Ridley piles** are the piles consisting of a reinforced concrete bar and a concrete cast-in-situ casing (Figure 2.23).

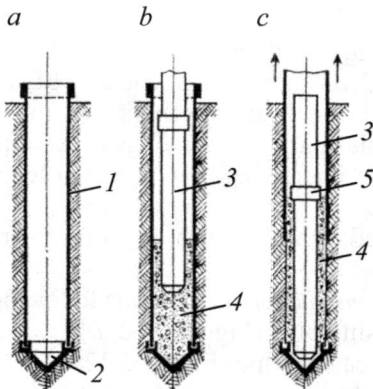

Figure 2.23. Process flow diagram of constructing Ridley piles:
*1* – casing pipe; *2* – shoe with coupling; *3* – reinforced concrete bar;
*4* – concrete mix;
*5* – bar enlargement

*The technology of Ridley pile construction* includes the following:

1. Driving a thick-walled casing pipe with a cast shoe and a coupling into the ground (Figure 2.23, *a*).

2. Supply of cast concrete to a casing; pressing of a reinforced concrete bar into a borehole filled with concrete mix (Figure 2.23, *b*).

3. Withdrawal of the pipe; lowering of the bar onto the shoe, which is accompanied by spreading concrete mix over the borehole walls (Figure 2.23, *c*). An enlargement is provided on the bar to prevent concrete mix from going out of the borehole.

The **MacArthur piles** without shell are formed by driving steel pipes 35 cm in diameter together with a tip into soil. The concreting is carried out with the concurrent withdrawal of the pipe. A mandrel with a hammer is lowered onto concrete mix, which provides the concreting density and the required contact of concrete with adjacent soil. The base enlargement is achieved by tamping of concrete mix. Such piles are made with a length up to 18 m.

**Driven cast-in-situ piles.** Boreholes for driven cast-in-situ piles are formed by driving into the ground a thick-walled casing pipe, which is freely supported by a cast-iron shoe.

After the casing pipe has been driven to the design level, a hammer is raised and a reinforcing cage is lowered into the pipe, and the pile is concreted.

Concrete mix is delivered in buckets with a capacity of 0.4 to 0.5 m$^3$ and loaded into the pipe to make the whole pile at a time or in two or three stages. Upon loading of every batch, the pipe is withdrawn to a certain height under the action of frequent hammer blows. The alternating hammer blows, upward along the pulling structure connected to the pipe and downward along the pipe, contribute to the pipe withdrawal.

Due to the upward hammer blow, the pulling depth of the pipe is approximately 1.5 to 2.0 times higher than its penetration depth resulted from the next downward blow. The downward blows are transferred through the pipe to concrete mix that goes out of the pipe into the borehole under gravity. When tamping concrete mix, a wavy side surface of the pile is formed that increases the pile bearing capacity. In some cases, it exceeds the bearing capacity of common driven reinforced concrete piles.

When withdrawing the pipe, its lower end shall be constantly buried by 2 m into concrete mix.

Figure 2.24 shows the technological sequence of constructing the driven cast-in-situ piles.

**Vibrex and Super Vibrex piles.** The pile construction using the Vibrex and Super Vibrex technologies includes the driving of a reusable casing pipe and its withdrawal by means of a vibrator. A borehole for a pile is made by means of the soil displacement when driving the reusable casing pipe, whose lower end is closed with a flat shoe to be lost in the ground.

*The technology of Vibrex pile construction* includes the following:

1. Moving of the rig to a planned pile location. Installation of the lost flat shoe. Watertight connection of the shoe to the pipe (Figure. 2.25, *a*).

2. Using a diesel-driven or hydraulic hammer, lowering of the pipe with the attached shoe to the design method (Figure 2.25, *b*).

3. Lowering of the reinforcing cage into the casing pipe (Figure 2.25, *c*).

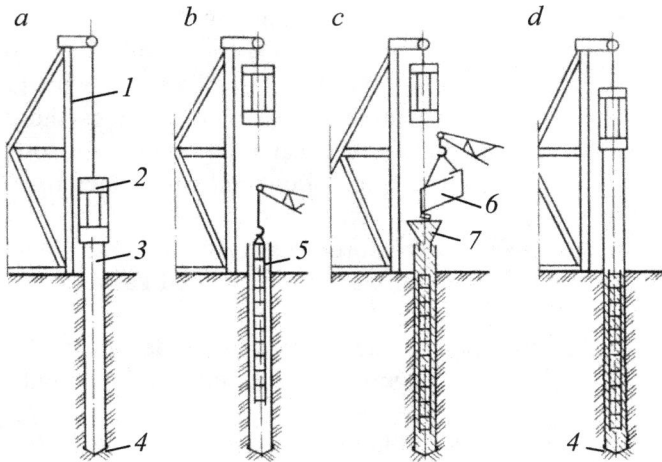

Figure 2.24. Process flow diagram of constructing driven cast-in-situ piles:
*a* – driving of pipe with lost shoe into ground; *b* – lowering of reinforcing
cage; *c* – concreting of pile; *d* – withdrawal of pipe with concurrently tamping concrete mix; *1* – pile driver; *2* – hammer; *3* – casing pipe; *4* – cast-iron
shoe; *5* – reinforcing cage; *6* – bucket; *7* – funnel

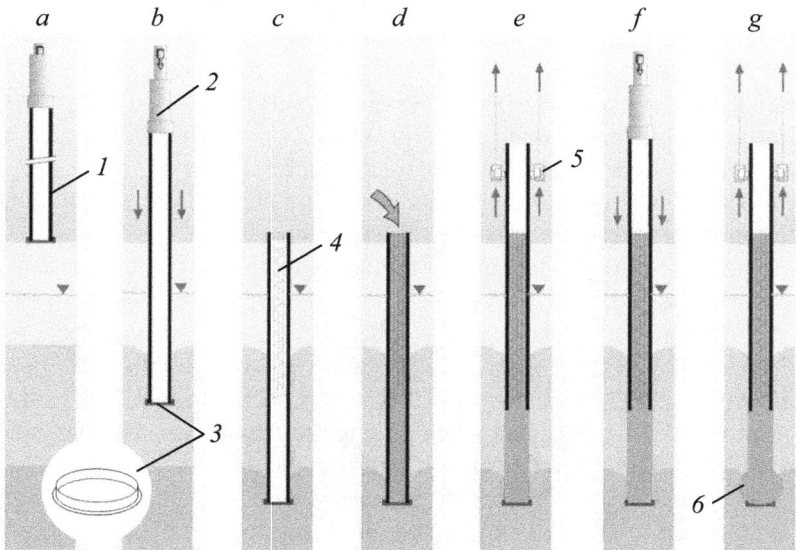

Figure 2.25. Technological operations to construct Vibrex (*a – e*)
and Super Vibrex (*a – g*): *1* – casing pipe; *2* – hammer; *3* – flat shoe;
*4* – reinforcing cage; *5* – vibrator; *6* – enlarged pile foot

4. Filling of the casing pipe with concrete mix (Figure 2.25, *d*).

5. Clamping of the free upper section of the pipe with a clamping mechanism of the hydraulic ring vibrator secured to the rig mast. Smooth withdrawal of the pipe from the ground by means of the vibrators with concurrently tamping concrete mix under the action of the pipe vibration (Figure 2.25, *e*).

*The Super Vibrex technology* (as opposed to the Vibrex technology) provides an enlargement of the pile foot, with the enlargement diameter being usually 1.5 to 3.0 diameters of the pile shaft. *The technology of Super Vibrex pile construction* includes the following:

1, 2, 3, 4. These operations are similar to the first four operations of the Vibrex pile construction (see Figure 2.25, *a – d*).

5. Clamping of the free upper section of the pipe with a clamping mechanism of the hydraulic ring vibrator secured to the rig mast; vibratory withdrawal of the pipe to a depth up to two meters from the previously reached design level of the pile tip (see Figure 2.25, *e*).

6. The repeated additional driving of the pipe upper section with the hammer; the enlargement of the pile base is formed as a result of compacted concrete mix (see Figure 2.25, *f*).

7. Final withdrawal of the pipe using the vibrator (see Figure 2.25, *g*).

To construct the Vibrex and super Vibrex piles, Fundex piling rigs (F12S, F12SE, F15, F16, F2800, F3500) are used that are equipped with hammers and special-purpose vibrators with ring clamps for the casing pipe. Such rigs can manufacture vertical and inclined piles with an incline of 4:1 from itself (forward) and 3:1 under itself (backward) at the same location where the rig is positioned. The technical features of the piling rigs and equipment to manufacture the Vibrex and Super Vibrex piles are given in Appendix 8. The maximum pile length is 37 m. The shaft diameter can be 273, 324, 356, 406, 456, 508, 556, 610 and 711 mm (Table 2.2).

*Table 2.2*

**Standard shoe diameters for Vibrex and Super Vibrex piles**

| Pile shaft diameter, mm | Diameter shoe, mm | Pile shaft diameter, mm | Diameter shoe, mm |
|---|---|---|---|
| 273 | 324 | 508 | 558 |
| 324 | 356 | 558 | 610 |
| 356 | 406 | 610 | 660 |
| 406 | 457 | 711 | 761 |
| 457 | 508 | – | – |

Among the advantages of the Vibrex and Super Vibrex technologies are the following:

- high bearing capacity of piles;
- high-quality forming of pile shaft due to compaction of concrete mix during vibratory withdrawal;
- there is no mud and it is not required to remove soil;
- it is possible to construct inclined piles;
- high production capacity.

The Vibrex and Super Vibrex technologies are used on the construction sites where it is allowed operating percussive machines.

## Piles in boreholes made by driving of shells to be lost in ground

The **Stern piles** are the prototype of a number of piles that are concreted in a lost metal shell driven into the ground.

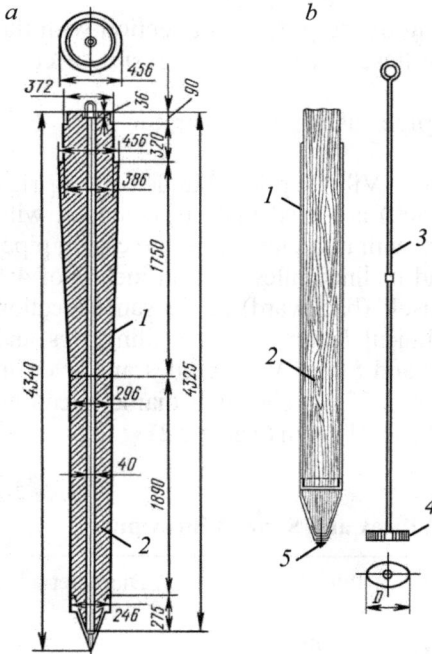

Prior to driving it into the ground, a wooden core is inserted into the shell, with a longitudinal through hole being provided in the core (Figure 2.26, *a*). A metal rod 40 mm in diameter with an anchor on its end is inserted into the hole. After the shell with the core has been driven to the design level, the core is extracted from the shell using the metal rod. Then the shell is filled up with concrete mix.

The **Mast piles** are a modification of the Stern piles and have a reinforced tip on the shell (Figure 2.26, *b*).

The advantages of the piles made in the lost shells are the guaranteed integrity of the pile shaft, the insulation of the shaft against corrosive water, the possibility to concrete piles in the considerable time after the shell has been driven down.

Figure 2.26. Designs of Stern (*a*) and Mast piles (*b*): *1* – metal shell; *2* – wooden core; *3* – metal rod; *4* – anchor; *5* – reinforced tip

Among the disadvantages of the Stern piles are the collapse of a shell and their insufficient rigidity, due to which problems arise to extract the wooden core. At the present time, the Stern and Mast piles are not used.

The **Peerless piles** include a cast metal shoe and reinforced concrete rings filled with concrete mix (Figure 2.27). The rings are lowered into the ground at the time of driving the core, which is a thick-walled steel pipe. After the rings have been lowered to the required depth, the core is withdrawn and the internal hole is filled with concrete mix.

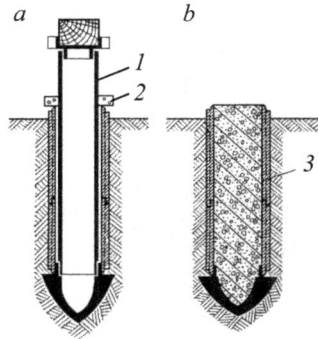

Figure 2.27. Peerless pile design:
*1* – core; *2* – coupling;
*3* – concrete ring

The advantage of the Peerless pile is higher bearing capacity as compared to the piles in metal shells. It is due to the fact that the friction against soil is higher for the concrete shell than for the metal shell.

The **Raymond piles** were designed in Poland in 1901. They include a tapered metal shell, which is driven into the ground and filled with concrete mix (Figure 2.28).

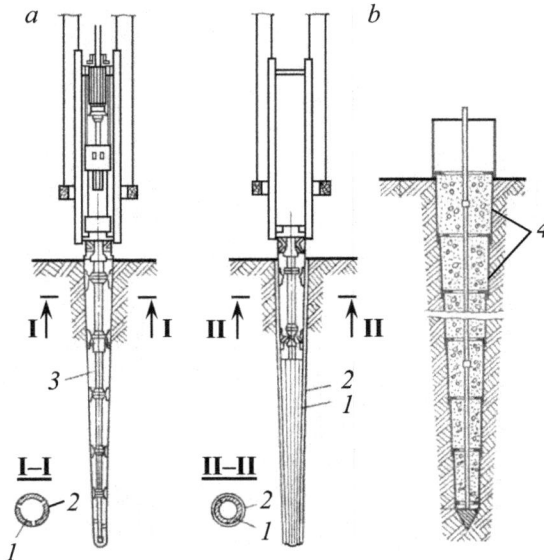

Figure 2.28. Design of core (*a*) and Raymond pile (*b*):
*1* – core shell; *2* – pile shell; *3* – core bar; *4* – sections of pile shell

The metal shell of the pile consists of multiple sections that are inserted into each other. The shell sections are made of sheet steel 1.0 to 1.5 mm thick. Along the outside surface of the sections, a spiral winding is welded made of six-millimeter wire to increase the shell rigidity.

The shell is driven into the ground by means of a hollow tapered core. The core is a shell split lengthwise in three parts. Into the core, a tapered bar is inserted that has wedge-shaped bosses and prongs. When a hammer blows the bar, the bosses and prongs push apart the core shell, due to which it is born against the pile shell. Upon reaching the design depth, the bar is removed, the split shell sections are shifted. Then the core is removed from the pile shell, and the pile is concreted.

### Piles in boreholes made by driving of reusable pipes, whose lower end is closed with concrete plug

**Franki piles.** For the construction of boreholes for piles, reusable casing pipes currently find wide application. When driven, the lower end of the reusable casing pipes is closed with a lost cast-iron or reinforced concrete shoe. When the pipe is raised, the shoe remains in the ground, forming a pile foot. In case of the Franki piles, a concrete plug that fills the lower end of the casing pipe serves as a shoe. The casing pipe is driven into the ground by

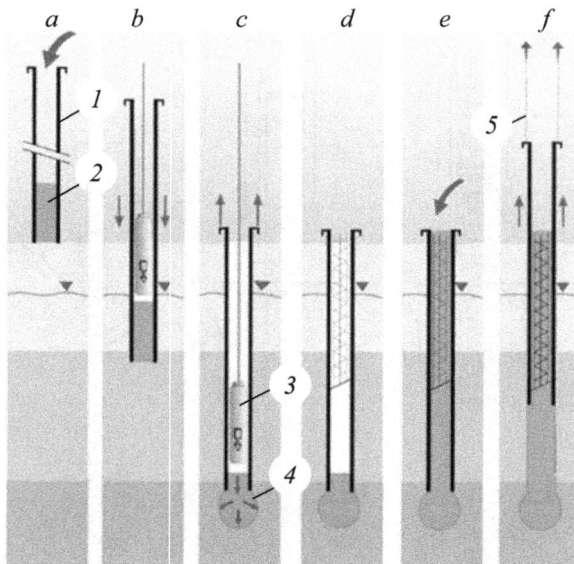

Figure 2.29. Process flow diagram of constructing Franki piles: *1* – casing pipe; *2* – concrete plug; *3* – tamper; *4* – enlarged pile foot; *5* – wire rope

hard blows on the concrete plug using a special-purpose tamper. At the design level, the casing pipe is held with wire ropes to prevent further penetration, and the plug is partially driven out into the ground, forming a bulb-shaped foot 1.5 to 2.0 times the pipe diameter.

To make the piles, there are used pipes 248 to 600 mm in diameter with the wall thickness of 32 mm.

*The technology of the Franki pile* construction is as follows (Figure 2.29). A casing pipe is positioned at the planned pile location, and approximately 200 l of dry concrete is poured into it. Concrete mix is tamped by blows of a tamper on a concrete plug. The concrete plug, becoming stuck in the pipe, pulls the pipe behind itself at the subsequent hammer blows and drives it into the ground.

Upon reaching the design depth, the casing pipe is pulled up with wire ropes. Then an additional batch of dry concrete is poured down the pipe and the plug is driven out by tamper blows. Being tamped into the ground, the plug forms an enlarged pile foot. Then a reinforcing cage is lowered into the casing pipe.

The shaft is concreted together with the removal of the casing pipe.

## 2.2.2. Piles in boreholes made by vibration displacement

Vibro piles designed by E. M. Perley and A. M. Rukavtsov started to be used in the domectic construction applications since 1960.

*The technology of constructing vibro piles with enlarged foot* includes the following operations:

1. Vibro-driving of a reusable steel pipe closed at the bottom with a lost shoe (Figure 2.30, *a*).

2. Filling of the pipe with cast concrete mix to a depth of 0,8 to 1,0 m (Figure 2.30, *b*).

3, 4. Forming of an enlarged foot using a tamper connected to the vibratory pile driver (Figure 2.30, *c*, *d*).

5. Lowering of a reinforcing cage (Figure 2.30, *e*).

6. Filling of the pipe with concrete mix followed by the pipe withdrawal using the vi-

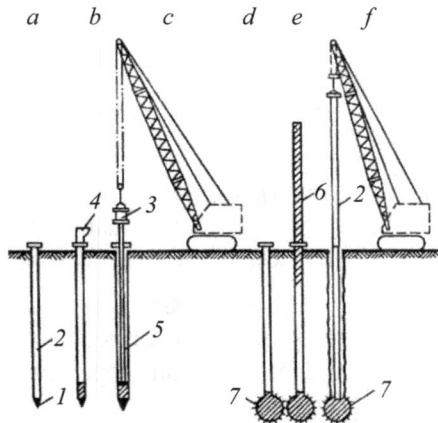

Figure 2.30. Process flow diagram for constructing vibro piles with enlarged foot:
*1* – lost reinforced concrete shoe;
*2* – reusable steel pipe; *3* – vibratory pile driver; *4* – funnel; *5* – tubulat tamper;
*6* – reinforcing cage; *7* – enlarged foot

bratory pile driver suspended on a boom of a self-propelled crane (Figure 2.30, *f*).

The vibration makes it possible to use dry concrete mix that provides the high quality of the pile shaft and reduces the concrete consumption.

To construct the vibro piles, there can be used multifunctional rigs such as Liebherr provided with Piling and Vibro Equipment vibratory pile drivers (Appendix 7).

### 2.2.3. Piles in reamed boreholes

**Piles made using helical tools (drilled displacement piles)** Boreholes for these piles are made without removing soil due to its horizontal displacement by means of a tool (Figure 2.31), which is a variable-section cone or cylinder with a screw blade. The tool (soil displacement auger – SDA) is attached to the end of a drill pipe, whose diameter is smaller than the maximum diameter of the tool.

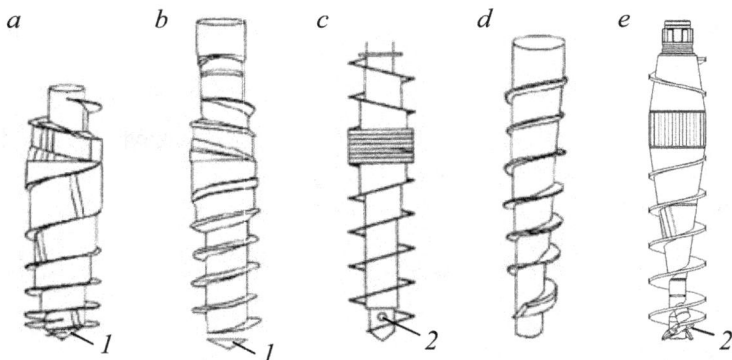

Figure 2.31. Helical tools to ream boreholes for drilled displacement piles:
*a* – Omega; *b* – De Waal; *c* – Berkel; *d* – Screwsol; *e* – Bauer;
*1* – lost shoe; *2* – hole to feed fine-aggregate concrete mix

When the tool penetrates into soil, the soil is displaced laterally and a densified body of soil is formed around the borehole, with the size of the body depending on the soil properties, the penetration speed and the tool design.

The pile concreting starts when the drill pipe with the tool reaches the design level. For this purpose, flowing fine-aggregate concrete mix is delivered into the drill pipe, which comes out through a hole in the tool, while the tool is being screwed out, and fills up the borehole.

The method to make boreholes using the helical tools was suggested in the Soviet Union by V. I. Feklin [51-54]. Later, the similar design efforts were developed and introduced into production by foreign companies [56-57].

*Bauer piles.* The Bauer pile construction technology has been developed in Germany. The Bauer rotary drilling rigs (Figure 2.32) and the RTG Rammtechnik helical tools (reamers) (Figure 2.33) are used to construct piles. There is a hole provided on the tool tip to deliver concrete mix. When the tool is lowered, the hole is closed with a plug. When the tool is withdrawn, concrete mix, which is delivered under pressure into the drill pipe, pushes out the plug.

Figure 2.32. Bauer rig with RTG Rammtechnik helical tool:
*1* – helical tool; *2* – drill pipe; *3* – standard mast of drilling rig; *4* – rotary head of drill pipe; *5* – nipple to feed concrete mix into drill pipe; *6* – jib; *7* – undercarriage

Figure 2.33. General view of rigs provided with helical tools

RTG Rammtechnik produces reamers to construct the piles 360, 440, 510 and 620 mm. The parameters of the hole-making using the Bauer rotary drilling rigs and the RTG Rammtechnik reamers are shown in Appendix 4 (Table A.4.4).

*The technology of Bauer pile construction* includes the following:

1. Installation of the drilling rig at the designated location (Figure 2.34, *a*).

2. Screwing the drill pipe together with the tool into the soil until the design level is reached (Figure 2.34, *b*, *c*).

3. Extracting the drilling tool with concurrently filling up the borehole with concrete mix via the hole in the tool tip (Figure 2.34, *d*).

4. Lowering a reinforcing cage suspended on a boom of a crawler caterpillar crane (Figure. 2.34, *e*).

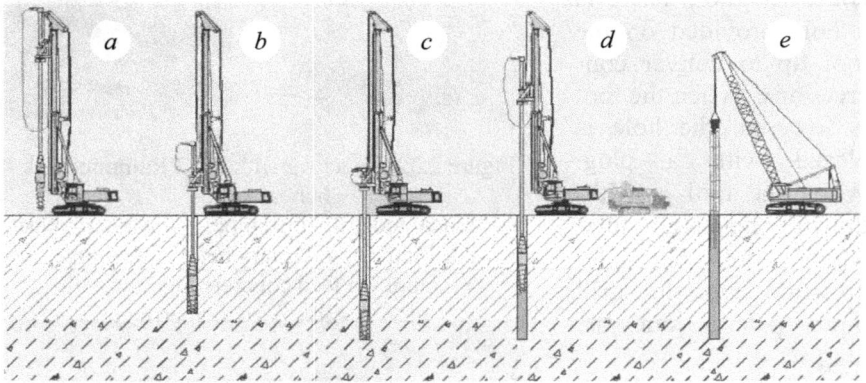

Figure 2.34. Process flow diagram of constructing displacement piles using helical tools

*Omega and De Waal piles* are constructed using helical tools, whose tip is closed with a lost shoe(see Figure 2.31). As opposed to a diameter of the hole in the RTG Rammtechnik reamer tip, a diameter of the lost shoe allows lowering the reinforcing cage into the borehole prior to extracting the tool,`that is, prior to concreting a pile. The tool designs have been developed in Belgium.

A tool to construct the Omega piles is a truncated cone 4 m long with a screw blade, whose spacing increases as its distances from the tip increases. Typically, the Omega piles are constructed using the Soilmec or Socofonda rotary drilling rigs.

The Omega and De Waal piles can be 310, 360, 410, 430, 460, 510, 560 and 610 mm in diameter. The Omega and De Waal piles can be as long as 30 m.

The discussed pile construction technologies using the helical tools have the following advantages:

- no considerable vibration and dynamic effect on adjacent buildings and structures;
- construction cycle is 20 to 25 min for the piles 550 mm in diameter and 25 m long;
- high quality of filling up a borehole with concrete mix due to its delivery under pressure (Figure 2.35);
- control of the drilling parameters with on-board computer;
- no mud when drilling.

Figure 2.35. Concreting of pile:
*1* – mixer truck; *2* – Soilmec concrete pump; *3* – Bauer drilling rig

Due to the soil consolidation around the borehole, the bearing capacity of the displacement piles is higher than the bearing capacity of drilled piles. Based on the data obtained by various analysts, the friction along the side`surface of those piles is approximately 30 % greater than for for the drilled piles, and the pile tip resistance is 50 to 70 % higher. In soft clay soils, the pile diameter increases due to the delivery of concrete mix into the borehole under pressure (Figure 2.36).

The technology of Bauer pile construction is widely used in Saint Petersburg. Despite of the decided advantages of the technology, it has a limited used as the construction of a pile field might result in the rising of the ground surface and the foundations of the buildings in the near vicinity.

**Atlas and Olivier piles.** The pile construction technologies have been developed in Belgium. When constructing the Atlas and Olivier piles, the boreholes for the piles are made without extracting soil due to its consolidation by means of a screwed-in reusable steel pipe with a cutting tip, whose lower end is closed by a shoe to be lost in ground (Figure 2.37). The inside diameter of the cutting tip is equal to the pipe diameter.

The first rigs to construct the Atlas piles made their appearance in the 60s of XX century (Figure 2.38) and were made by the Flemish Company, which later became known as Atlas Piling Company.

To construct the Atlas piles, the Atlas rigs (Figure 2.39) are used that are equipped with a hydraulic gear, which provides the simultaneous rotation and vertical movement (screwing-in and screwing-out) of a casing pipe with a cutting screw tip. The modern Atlas drilling rigs (BT-42, BT-60) are operated by two operators and have a capacity that allows constructing 200 running meters per 8-hour shift.

Figure 2.36. Piles with shaft enlargement formed in soft soil layer by delivering concrete mix under pressure

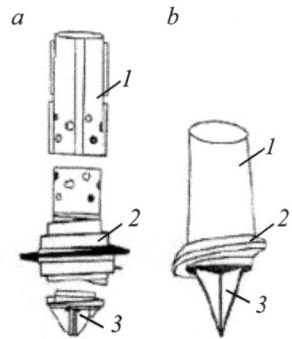

Figure 2.37. Screw tips used in construction of Atlas (*a*) and Olivier (*b*) piles: *1* – drill pipe; *2* – cutting tip; *3* – lost cast-iron shoe

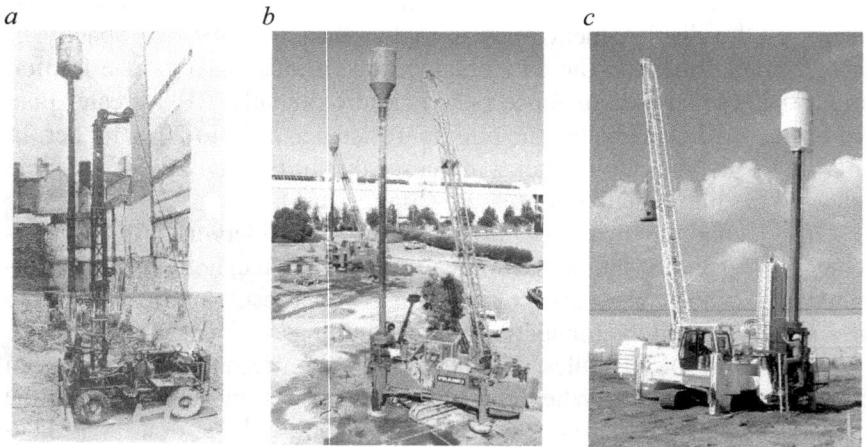

Figure 2.38. Atlas rigs: *a* – beginning of 1960s; *b* – 1980s; *c* – modern BT-60 Rig

66

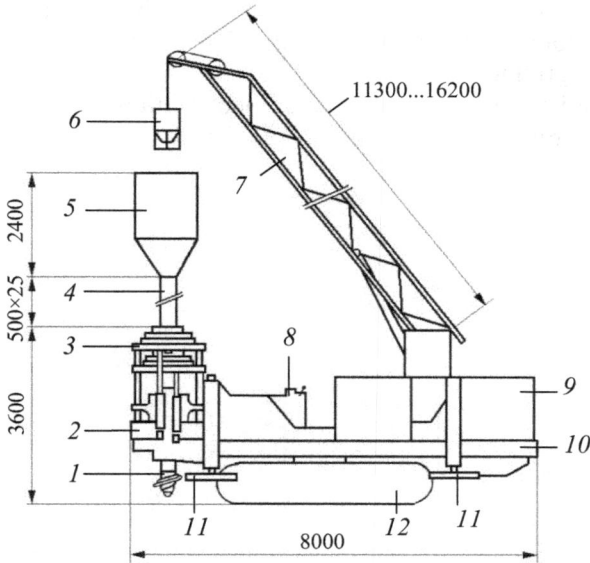

Figure 2.39. Atlas BT-40 rig: *1* – cutting screw tip; *2* – drilling table;
*3* – jacking mechanism; *4* – casing pipe; *5* – hopper; *6* – bucket to
fill hopper; *7* – lifting beam; *8* – control panel; *9* – engine; *10* – rig
platform; *11* – supports; *12* – caterpillar tracks

*The technology of the Atlas* and *Olivier pile construction* includes
the following:

1. Installation of the drilling rig at the drilling location (Figure 2.40, *a*). Sealing of a joint between the drill head and lost shoe with watertight ductile material.

2. Non-vibration clockwise screwing of the pipe and the drill head closed underneath with the lost shoe (Figure 2.40, *b*) into the ground under the action of torsion torque and vertical force. Using the on-board computer of the drilling rig, recording of the force transferred to the screwed-in pipe; the rotation speed and the pipe screwing time; the concrete volume placed in the borehole.

3. Lowering of the reinforcing cage in the pipe after the required depth as been reached (Figure 2.40, *c*).

4. Filling up of the borehole with concrete mix through the casing pipe(Figure 2.40, *d*).

5. Counterclockwise screwing out of the pipe together with the cutting tip (Figure 2.40, *e*). As this takes place, soil is pressed out again, and concrete mix gradually fills up resulting voids under the action of hydrostatic pressure. In order to prevent the pile from being compressed by soil,

the hydrostatic pressure of concrete mix at the base of the casing pipe (borehole bottom) shall be considerably higher than the joint pressure of soil and underground water.

6. If required, lowering of an additional reinforcing cage into the upper part of the pile and forming of the pile head.

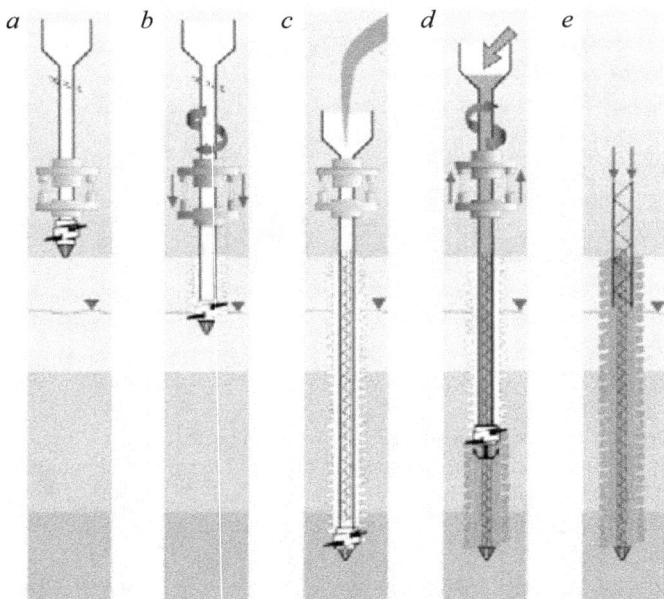

Figure 2.40. Process flow diagram of constructing Atlas piles

The Atlas cutting tips are produced 310, 360, 410, 460 and 510 mm in diameter with the screw blades 460, 510, 560, 610 and 660 mm in diameter, respectively. An Atlas pile can be as long as 24 m.

Figure 2.41 shows the Atlas pile heads with the typical screw shape of the shaft. Usually, the width of the pile blade is not greater than 50 mm.

For the construction of the Olivier piles, there are used cutting tips 310, 360, 410, 460 and 510 mm in diameter with the screw blades 460, 510, 610, 660 and 710 mm in diameter, respectively.

The Atlas and Olivier technologies have a number of advantages:
- works can be carried out in the vicinity of existing buildings as there is low-level noise and no vibration;
- an increase in the bearing capacity of a pile due to the soil consolidation by the tip;
- a wide variety of the tips, which allows operating in almost all types of dispersed soils;
- high production capacity.

Figure 2.41. Atlas pile heads

**Fundex piles.** The Fundex pile construction technology has been developed in the Netherlands. Boreholes for the piles are made without extracting soil due to its consolidation by means of a screwed-in reusable steel pipe, whose lower end is closed by a screw tip to be lost in ground. The Fundex piles are constructed using multifunctional Fundex rigs (F12S, F12SE, F15, F16, F2800, F3500), whose technical features are given in Appendix 8.

A lost screw cast-iron tip of various designs serves as a foot of the pile to be constructed (Figures 2.42, 2.43).

The tip is set at the specified point on the surface of the soil base. The lower end of the reusable thick-walled pipe is fastened to the tip by means of a bayonet connection through a waterproof soft pad; the upper end of the pipe is clamped in the power working attachment of the drilling table that moves along the guide mast of the rig.

The borehole for the pile to be constructed is made by the rotating-and-pushing penetration of the drill pipe with the lost shoe down to the specified level (Figure 2.44). While the drilling tool advances into the soil base, soil is pushed apart radially from the borehole centerline and consolidated at the same time. When the tip reaches the design level, the pipe is checked for the absence of water.

Figure 2.42. Lost screw tips
for Fundex piles

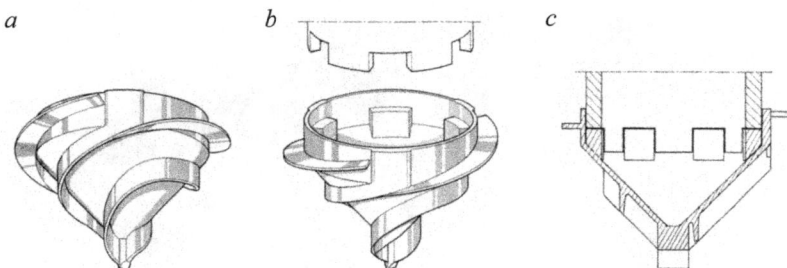

Figure 2.43. Lost screw tip: *a* – general view; *b* – connection of tip and pipe connector; *c* – cross section of connection

A reinforcing cage is lowered into the dry reusable pipe through its open upper end. Prior to delivering concrete mix, a batch of primer including cement, sand and water (1:1:1) is fed into the pipe to prevent concrete mix from disintegrating. Then the pipe is batch-filled with fine-aggregate (5 to 20 mm) plastic concrete with cone slump of 12 to 14 cm. The drill pipe is extracted from the ground by backward rotation with concurrently pulling it.

Tips 450, 560 and 660 mm in diameter are produced to construct the piles 380, 460 and 540 mm in diameter, respectively. A pile can be as long as 40 m.

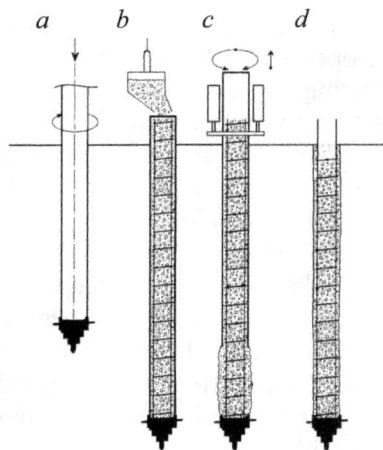

Figure 2.44. Process flow diagram of constructing Fundex piles

The Fundex technology has been used worldwide since 1960s, and in Russia, it was first used in Saint Petersburg in 2001.

Among the advantages of the Fundex piles are the following:
- no use of prefabricated piles and no operations associated with prefabricated piles (delivery, store-keeping, lifting to pile driver, jointing, etc.);
- no considerable dynamic effect on soil base in the course of pile construction, which is especially important when carrying out works in restricted conditions of existing development;
- no works are required to remove soil from the drill pipe and to transport it from construction site;
- high production capacity (up to 380 running meters/day);
- low-level of noise when the drilling rig is operating.

**Tubex piles** consist of a metal shell and a reinforced concrete shaft. The Tubex pile construction technology includes the screwing-in of a thick-walled metal casing pipe with a tip down to the required depth, the lowering of a reinforcing cage into the pipe and the filling of the pipe with concrete mix.

As opposed to the Fundex technology, no provision has been made in the Tubex technology to extract the casing pipe from the soil base. Therefore the Tubex technology is used when constructing piles in soils with the seepage flow rate higher than 200 m/day and for the stabilization of current landslide slopes.

As it has pressure-tight connection to the tip, the lost casing pipe prevents the borehole walls from the collapsing and the shaft from narrowing, providing the high-quality pile construction.

A disadvantage of the Tubex piles is smaller, as opposed to concrete piles, friction on soil, that is, a lower soil-specific bearing capacity of the piles.

For the construction of the Tubex piles, screw tips are used that have a design different from the Fundex tip design. The size of the standard Tubex tips is given in Table 2.3. A Tubex pile can be as long as 30 m.

*Table 2.3*

**Diameters of standard screw tips for Tubex and Tubex Grout-Injection piles**

| Tubex piles | | Tubex Grout-Injection piles | |
|---|---|---|---|
| Screw tip diameter, mm | Diameter of casing pipe, mm | Screw tip diameter, mm | Diameter of casing pipe, mm |
| 310 | 168, 219 | 300 | 220 |
| 450 | 324, 355 | 450 | 324 |
| 560 | 355, 406, 457 | 560 | 355 |
| 620 | 406, 457 | 560 | 368 |
| 670 | 457, 508 | 560 | 406 |
| 850 | 609 | 670 | 457 |
| 950 | 762 | 670 | 508 |

**Tubex Grout-Injection (TGI) piles** To construct a pile, a thick-walled casing pipe is screwed into the ground, with a screw tip being welded to the casing pipe. In the tip, there are provided holes, through which cement grout is injected while the pipe is being screwed in.

To deliver cement grout into the casing annulus, the holes in the tip are connected to a metal injection tube placed inside the casing pipe. As the maximum diameter of the screw tip is smaller then the outside diameter of the casing pipe, cement grout penetrating into the casing annulus forms "a jacket" around the pipe (Figure 2.45).

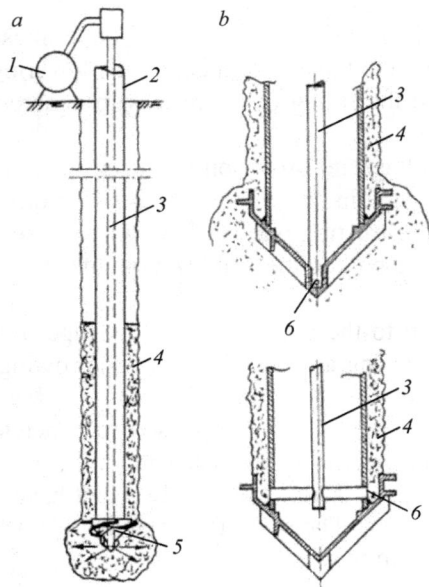

Figure 2.45. Forming of borehole for Tubex Grout-Injection pile (*a*) and cross sections of screw tips (*b*): *1* – concrete pump; *2* – casing pipe; *3* – injection tube; *4* – cement "jacket"; *5* – screw tip; *6* – hole to inject cement grout

Figure 2.46. Fundex F3500 Rig

In the course of borehole-making, cement grout cools down the pipe with the tip and reduces their friction on soil.

After the pipe has been lowered to the required depth, the injection tube is withdrawn out of it. Then a reinforcing cage is lowered into the casing, and concrete mix is poured.

When construction the piles inside structures, the casing pipe is welded of multiple sections.

The size of the standard tips used for the construction of the Tubex Grout-Injection piles is given in Table 2.3.

Due to the cement crust forming around the casing pipe, the bearing capacity is higher for the Tubex Grout-Injection piles than for the Tubex piles in metal shell.

The boreholes for the Fundex, Tubex and Tubex Grout-Injection piles are made using multifunctional Fundex rigs (F10, F12S, F12SE, F15, F16, F2800, F3500). When reconstructing structures, the Tubex and Tubex Grout-Injection piles are made using small-scale Tubex rigs (RBM35, TBM12, TBM35, TD35, TBX35).

# Chapter 3. PILES USED IN UNDERPINNING

To increase the bearing capacity of the foundations for the constructed buildings, piles of small cross section are used that are fabricated in factory or constructed directly in the ground. The size of the faces or the diameter of those piles does not exceed 250 to 300 mm.

Abroad, the piles concreted in the drilled boreholes no more than 250 mm in diameter are called *micropiles*. Depending on the technology of the pile shaft formation, there are four types of the micropiles: *A*, *B*, *C* and *D*.

*Type A micropiles* are constructed by way of filling up drilled boreholes with cement or cement-and-sand grout from top downward under gravity. As the grout-filled borehole is not pressured up, the shaft diameter does not expand. Such piles are usually used to take up compressive loads.

To construct the *Type B micropiles*, cement grout is used that is delivered into the borehole from bottom upward through an injection tube under the pressure of 0.5 to 1.0 MPa (Figure 3.1, *c*). The pressure of 0.5 to 1.0 MPa prevents the bohehole walls from collapsing and soil from being hydraulically fractured.

The *Type C micropiles* are constructed in two stages. At the first stage, cement grout is poured into a borehole as in the case of constructing the Type *A* micropiles. The second stage is started after 15 to 25 min after the first stage has been completed and before cement grout has begun to set. At the second stage, cement grout is introduced into the borehole from bottom upward through an injection tube under the pressure of 1.0 MPa, with the borehole mouth being plugged (Figure 3.1, *d*). The technology of the Type *C* micropile construction is also referred to as IGU (injection global and unitary).

As with the *Type C micropiles*, the *Type D micropiles* are also constructed in two stages. During the first stage, cement grout is poured into a borehole from top downward as in the case of constructing the Type *A* micropiles. Cement grout can be pressured up with low pressure as in the case of constructing the Type *B* micropiles. At the second stage, cement grout is injected through a tube under the pressure of 2 to 8 MPa (Figure 3.1, *e*). An injection tube with holes is used for the selective injection at specified depths. The holes in the tube are covered with collars. A packer – an attachment for covering and sealing individual section of the tube – travels inside the tube. This method of the pile construction is used worldwide and called IRS (injection repetitive and selective).

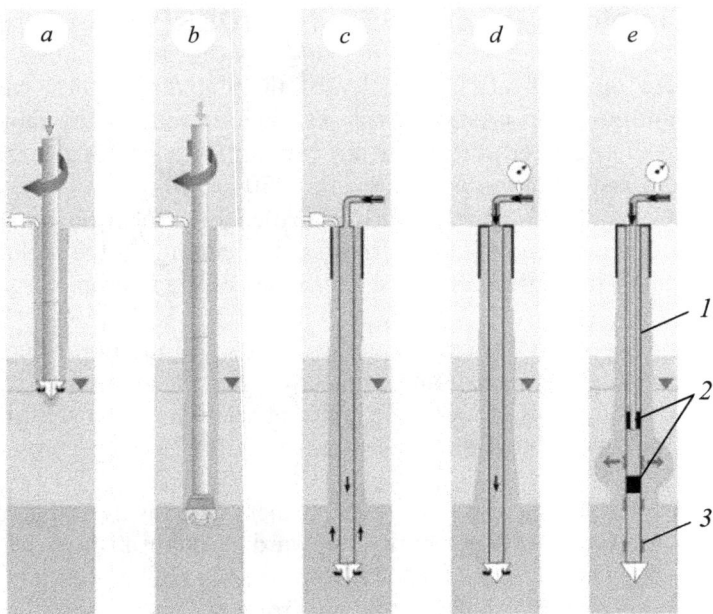

Figure 3.1. Some technological operations of micropile construction:
*a* – mud-circulating drilling of borehole by means of drill bit; *b* – mud-circulating drilling of borehole by means of three-cone bit; *c* – filing up of borehole with cement grout from bottom upward through injection tube under low pressure; *d* – injection global and unitary under low pressure; *e* – repetitive selective injection under high pressure; *1* – injection tube with holes; *2* – packer; *3* – collar

## 3.1. Bored injected piles

In many instances, foundations to be underpinned in the existing buildings and foundations are used as a raft for a new foundation constructed on bored injected piles (Figure 3.2). When underpinning the strip stonework foundation, usually the protective grouting of stone masonry comes before the construction of bored injected piles.

For the purpose of the protective grouting, the injected piles are drilled using core rigs with air flushing. A borehole diameter is established depending on the working conditions, the masonry conditions and the dimensions of the existing foundation. When underpinning foundations, the grouting is carrying out, as a rule, in two stages. At the first stage, a borehole is drilled within the limits of the foundation stopping 0.5 m above the foundation base. In the borehole mouth, a plug (obturator) is installed to prevent the injected grout from going out of the borehole, and then the foundation masonry is grouted. Upon completion of grouting, the borehole is aged for 2 to 3 days.

At the second stage, the foundation body is drilled over again down to its base and further to a depth of 0.4 to 0.5 m, and the foundation-ground contact area is grouted. In this case, the plug is placed in the foundation masonry 0.5 m above the base.

The injection pressure is not higher than 0.1 MPa in grouting the foundation masonry and 0.2 MPa in grouting the contact area. The injection is stopped if the grout flow rate does not exceed 1 l/min under the pressure of 0.2 MPa during 10 min.

For grouting, there is used cement grout, the type and composition of which depend on the design, material and conditions of the existing foundations, the geological and hydrogeological conditions of the site. In every particular case, the grout composition is selected in an laboratory.

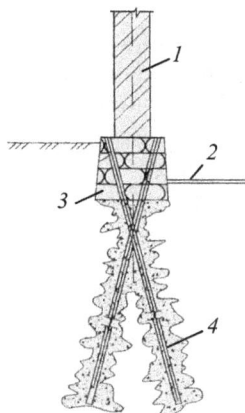

Figure 3.2. Example of underpinning of strip foundation with bored injected piles:
*1* – wall; *2* – basement floor;
*3* – foundation to be underpinned;
*4* – bored injected pile

*The technology of constructing bored injection piles for the foundation underpinning* lies in the following. In a foundation, a borehole is drilled to install a conductor, whose inside diameter is greater than or equal to the design diameter of a bored injection pile. Due to the small pile diameter, the inclination of the borehole to the vertical axis does not exceed 15°. The borehole is made using core rigs with air flushing. The borehole for the conductor is filled up with cement-and-sand grout until it spills out of the borehole mouth. When the grout level reduces in the borehole by more than 1 m, the borehole is aged for a day and then refilled up to the mouth with cement-and-sand grout of lower W/C. A conductor is installed into the borehole before cement grout has begun to set. In two days, cement stone is drilled out in the conductor with air flushing. Then the borehole is drilled in the ground down to the design level of the pile bottom end.

To avoid excessive soil excavation when drilling the borehole, there are used continuous flight augers with blades of small width. In some cases, when making boreholes in unstable watered soils, the drilling is carried out with flushing the boreholes with drilling (bentonite) mud or using casing pipes.

Upon completion of drilling, sludge is flushed out of the borehole with drilling mud for 3 to 5 min.

A reinforcing cage is lowered into the borehole as individual sections before (Figure 3.3) or after the pile has been concreted. The cage sections are jointed by welding.

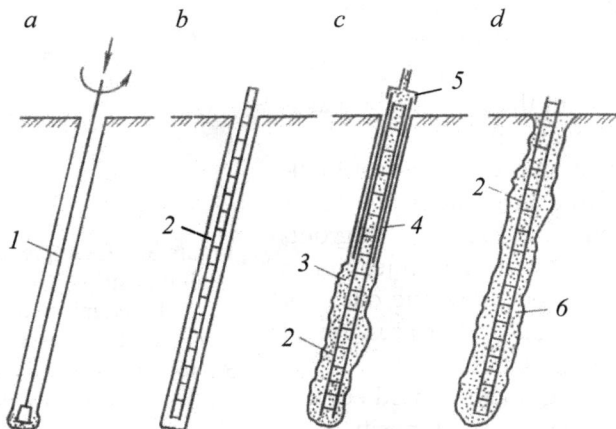

Figure 3.3. Process flow diagram of constructing bored injection piles:
*a* – drilling of borehole; *b* – lowering of reinforcing cage into borehole;
*c* – filling of borehole with grout, pressuring-up; *d* – finished pile; *1* – working attachment of drilling rig; *2* – reinforcing cage; *3* – cement grout;
*4* – conductor; *5* – obturator; *6* – cement stone

After the reinforcing cages have been installed in the design position, grout is pressured up in the borehole if there is no grout leakage from the borehole (the reduction of the grout level no more than by 0,5 m in the borehole). To that effect, a plug (obturator) with a pressure gauge is in-stalled in the upper part of the conductor, and grout is injected through an injector under the pressure of 0.2 to 0.3 MPa for 3 to 4 min. The pressur-ing-up can be stopped if the total grout flow rate is not higher than 200 l. In case of higher flow rate, it is required to age piles for a day, following which the pressuring-up is to be repeated.

For the most part, the bored injected piles are used in underpinning the foundations of reconstructed buildings and facilities. Sometimes they are used in construction of new buildings in vicinity of the existing ones.

To construct bored injection piles inside buildings, there is used small-sized equipment such as portable Sterh drilling rig or Alligator crawler drilling rig (Appendix 3). Overall dimensions of Alligator Rig: length 2.5 m, width 1.0 m; height 1.8 m.

**Soilex piles.** These piles were developed in Sweden in 80s of the last century. Their specific feature is an enlargement, which forms when cement or cement-and-sand grout is injected into a shell. The shell is a bundle of rolled metal sheet (Figure 3.4) placed on the borehole bottom. Upon completion of injection, an enlarged pile foot is formed, whose diameter 5 to 10 times the pile shaft diameter. Usually, the injection pressure is 0.5 to 3.0 MPa and depends primarily on the shell installation depth and the soil type.

Figure 3.4. General view of expanding bodies Soilex

Table 3.1 shows the standard sizes of the expanding Soilex shells.

*Table 3.1*

**Range of expanding bodies Soilex**

| Type of expander bodies | Before expansion | | | | After expansion | | | |
|---|---|---|---|---|---|---|---|---|
| | Length, m | Cross section, mm | Weight, kg | Minimum diameter of borehole, mm | Length, m | Diameter, m | Cross-sectional area, m$^2$ | Volume, l |
| EB 410 | 1.0 | 70×70 | 18 | 87 | 0.8 | 0.4 | 0.12 | 75 |
| EB 430 | 3.0 | 70×70 | 45 | 87 | 2.8 | 0.4 | 0.12 | 325 |
| EB 512 | 1.2 | 80×80 | 26 | 115 | 1.0 | 0.5 | 0.20 | 125 |
| EB 517 | 1.7 | 80×80 | 35 | 115 | 1.5 | 0.5 | 0.20 | 220 |
| EB 530 | 3.0 | 80×80 | 59 | 115 | 2.8 | 0.5 | 0.20 | 480 |
| EB 614 | 1.4 | 95×95 | 35 | 135 | 1.1 | 0.65 | 0.33 | 200 |
| EB 630 | 3.0 | 95×95 | 72 | 135 | 2.7 | 0.65 | 0.33 | 675 |
| EB 815 | 1.5 | 110×110 | 54 | 160 | 1.1 | 0.8 | 0.50 | 340 |
| EB 825 | 2.5 | 110×110 | 80 | 160 | 2.1 | 0.8 | 0.50 | 845 |
| EB 830 | 3.0 | 110×110 | 95 | 160 | 2.6 | 0.8 | 0.50 | 1100 |

The technology of forming the Soilex enlargement can be combined with constructing a CFA-pile or a pile protected by a casing pipe (Figure 3.5).

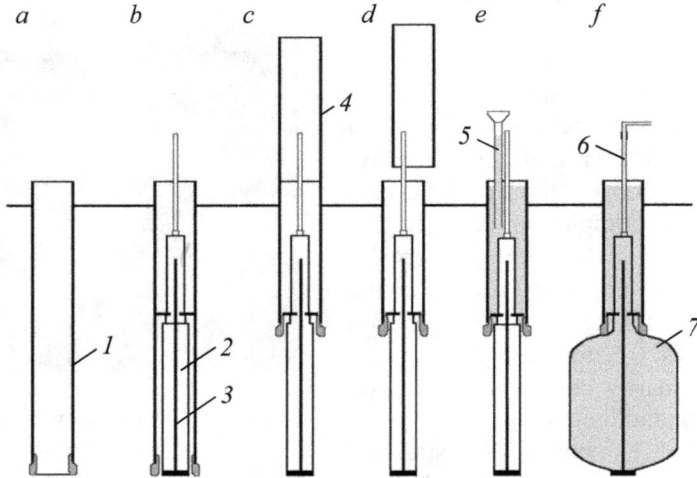

Figure 3.5. Process flow diagram of constructing Soilex
pile protected by casing pipe:
*a* – borehole with casing pipe; *b* – lowering of bundle of metal sheets;
*c* – withdrawing of casing pipe; *d* – removal of upper section of casing pipe;
*e* – concreting of pile shaft; *f* – delivery of cement grout into sheet bundle;
*1* – casing pipe; *2* – expanding shell; *3*– steel bar; *4* – section of casing pipe;
*5* – tremie pipe; *6* – injection tube; *7* – expander body

*The technology of Soilex CFA-pile construction* includes the following:
1. Drilling of a borehole with continuous flight auger.
2. Gradually withdrawing the auger while concurrently filling up the borehole with concrete mix.
3. Lowering of a bundle of metal sheets onto the borehole bottom.
4. Lowering of a reinforcing cage.
5. Injecting grout into the metal sheet bundle.

The expanding shell clamps soil surrounding the pile base, which increases its bearing capacity. For this reason, the pile length or diameter can be reduced, which allows using small-sized drilling equipment.

As the formed enlargement can take up both compressive and tensile loads, the Soilex piles can be used as load pulling forces, which allows using these piles as ground anchors

In sand soils, it is efficient to *construct Soilex piles using vibratory pile drivers*. The technology implies the vibratory penetration of a steel pipe and expanding shell welded to its end. In the course of penetration, the expanding shell is not deformed.

Generally, there are used steel pipes 10 to 15 m long and 0.15 to 0.25 m in diameter. To drive a casing pipe of such dimensions, small vibrators are sufficient and their use allows increasing the production capacity. For the casing pipe 15 m long, it takes usually few minutes to drive it into medium-density sand. The use of high-frequency vibratory pile drives allows reducing the risk of damages to surrounding buildings.

In compact soils and in boulder-containing soils, it is a good practice to use rotary drilling equipment in order to avoid damages to expanding shell.

The advantages of the Soilex piles lie in the following:

- no noise and vibration provides the safe use of the technology in restricted conditions;
- the "expanding body" technology can be combined with the other pile-penetration methods;
- the enlarged foot allows reducing the pile length;
- the bearing capacity of the Soilex piles can be determined by way measuring a volume of grout to be injected into the shell and the injection pressure.

**Bored injected Titan anchor piles.** These piles were developed in Germany in 1984. They consist of an anchor tie and cast-in-situ reinforced concrete shaft. The anchor tie (Figure 3.6) in the form of a continuous-thread pipe serves as drilling rod and injection tube at the same time. A range of the Titan anchor ties is given in Table 3.2.

The boreholes for the Titan anchor piles are drilled without casing pipes by the percussive-rotary drilling method, with delivering flushing cement grout of W/C 0.7 to 1.0 under the pressure of 1 to 2 MPa via the internal channel of the rod.

A borehole is drilled by means of a drill bit fastened to the rod end (Figure 3.7, *a*). A type of the drill bit and the grout composition are selected depending on the properties of soil to be drilled. In the drill bit, there are two side holes (nozzles) to deliver flushing grout via the drilling rod to the borehole bottom. When the drilling rod rotates, flushing grout performs the radial "cutting" of soil for a shaft of an anchor pile. Flushing grout is displaced to the borehole mouth and carries away the drilled soil.

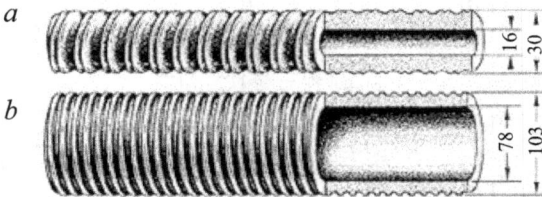

Figure 3.6. Detail of anchor tie:
*a* – TITAN 30/16; *b* – TITAN 103/78

Table 3.2

**Range of Titan anchor ties**

| Feature | TITAN 30/16 | TITAN 30/11 | TITAN 40/20 | TITAN 40/16 | TITAN 52/26 | TITAN 73/53 | TITAN 73/45 | TITAN 103/78 | TITAN 103/51 |
|---|---|---|---|---|---|---|---|---|---|
| Nominal outside diameter, mm | 30 | 30 | 40 | 40 | 52 | 73 | 73 | 103 | 103 |
| Effective outside diameter (for static analysis), mm | 27.2 | 26.2 | 36.4 | 37.1 | 48.8 | 69.9 | 70.0 | 100.4 | 98.0 |
| Inside diameter, mm | 16 | 11 | 20 | 16 | 26 | 53 | 45 | 78 | 51 |
| Permissible tensile-compressive load, kN | 100 | 150 | 240 | 300 | 400 | 554 | 675 | 1000 | 1500 |
| Permissible shear stress, kN | 58 | 88 | 138 | 164 | 240 | 329 | 390 | 578 | 899 |
| Ultimate load, kN | 220 | 320 | 539 | 660 | 929 | 1160 | 1630 | 2282 | 3460 |
| Weight, kg/m | 3.0 | 3.5 | 5.6 | 6.9 | 10.5 | 12.8 | 17.8 | 24.7 | 43.4 |
| Cross section, mm$^2$ | 382 | 446 | 726 | 879 | 1337 | 1631 | 2260 | 3146 | 5501 |
| Yield point, kN | 180 | 260 | 430 | 525 | 730 | 970 | 1180 | 1800 | 2726 |
| Yield stress $T_{0.2}$, MPa | 470 | 580 | 590 | 590 | 550 | 590 | 510 | 570 | 500 |
| Moment of inertia, cm$^4$ | 2.37 | 2.24 | 7.82 | 8.98 | 25.6 | 78.5 | 97.6 | 317.0 | 425.0 |
| Moment of resistance, cm$^3$ | 1.79 | 1.71 | 4.31 | 4.84 | 10.5 | 22.4 | 27.9 | 63.2 | 86.3 |
| Plastic moment of resistance, cm$^3$ | 2.67 | 2.78 | 6.70 | 7.83 | 16.44 | 32.1 | 41.9 | 89.6 | 135.0 |
| Permissible torsion torque, N·m | 487 | 649 | 1506 | 1784 | 3216 | 8202 | 8449 | 20940 | 24818 |
| Permissible impact energy, J | 84 | 84 | 145 | 145 | 400 | 610 | 610 | 610 | 610 |

*Notes*: 1. The permissible tensile-compressive loads make allowance for the safety factor of 1.75. 2. Permissible transverse force was determined with an allowance for plastic deformation. 3. Permissible torsion torque makes allowance for the safety factor of 2.0.

The borehole flushing results in forming a cement cake on the borehole walls, which prevents the walls from collapsing. Due to the infiltration of cement binding agent into the base, the soil resistance increases along the side surface of the pile.

After the borehole was drilled and flushed under the pressure of 2 to 6 MPa, it is injected with cement grout of $W/C = 0.4$ (Figure 3.7, $b$). The anchor root is formed when the drill bit moves back forth with the radial delivery of injection grout. The delivery of cement grout is continued until it flows out of the borehole mouth.

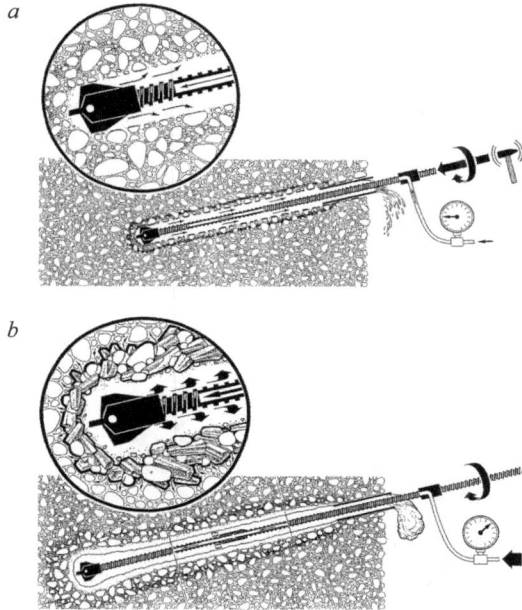

Figure 3.7. Technological operations in construction of bored injection Titan piles:
$a$ – percussive-rotary drilling with flushing;
$b$ – forming of pile shaft

The rod is centered in the borehole by means of braces screwed on the rod (Figure 3.8, $a$). Figure 3.8, $b$ shows the design of the anchor pile.

Upon completion of forming the pile body, a plastic pipe is put on the exposed end of the drilling rod, which serves as load-bearing element, to provide protection during the operating period.

According to the data provided by Ischebeck, the company that developed the Titan pile design, the diameter $D$ of the anchor root is at least as follows after injection:

- $2d$ in coarse and medium coarse sand ($d$ = diameter of drill bit);
- $1,5d$ in fine sand;
- $1,4d$ in cohesive soil (sand loam, loam, clay).

After cement stone has matured, piles undergo check tests by static load. If an anchor passes the check tests, its pretensioning is performed by way of screwing a ball nut.

The bored injection Titan piles are successfully used not only for the underpinning of foundations, but also for strengthening sheet-pile walls of foundation pits, reinforcing of slopes and constructing foundations of structures susceptible to tripping.

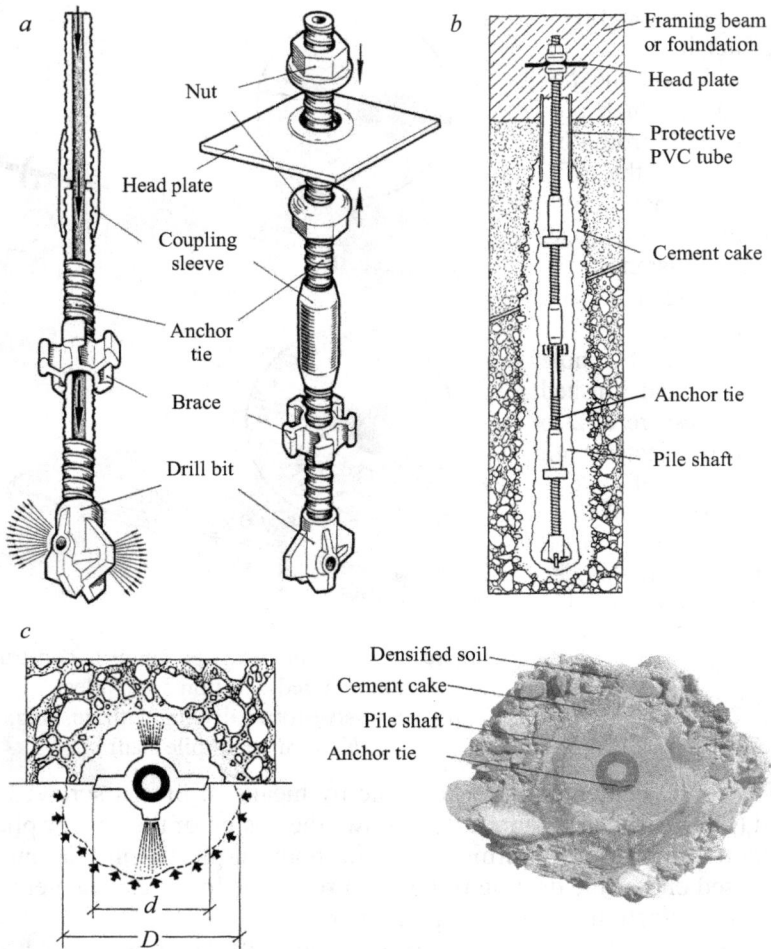

Figure 3.8. Design of drilling rod (*a*) and bored injection Titan pile (*b*); cross section of pile (*c*)

## 3.2. Multisection jacked piles

For the most part, such piles are used in underpinning the foundations of existing buildings and facilities. They allow unloading soils in the upper area of a base and transferring the load from a building to relatively firm soils.

The distinctive feature of the jacking technology of multisection (sectional) piles of small cross section lies in using the dead load of a reconstructed structure for the pile-driving instead of heavy-duty filling machinery.

The matters of management and work technique related to the under-pinning of foundations with jacked piles were studied in NIIpromstroy resulting in the compilation of VSN 16–84 (Instruction on underpinning of foundations of failure-condition and reconstructed buildings using multisection piles).

Mega piles are a modification of the multisection jacked piles. The piles are well known both in our country and abroad. Thus, the Mega piles were used to restore the majority of the old buildings in Helsinki and Turku that were deformed due to putrefaction of heads of timber piles.

**The Mega piles** are constructed of precast reinforced concrete elements 60 to 120 cm long with the cross section of 20×20, 25×25 and 30×30 cm. The elements are arranged over each other, being in contact with their end surfaces. To take up horizontal forces, vertical dowel wedges 37.5 to 50.0 mm in diameter are usually placed between the elements, with those dowel wedges preventing the elements to be joined from displacing and allowing their insignificant turn. The Mega piles with circular cross section are convenient to be rolled over in restricted basement conditions.

In NIIpromstroy, detailed drawings were made for multisection reinforced concrete piles of square cross section 30×30 cm and 0.6; 0.9 and 1.2 m long with non-tensioned reinforcement (Table 3.3). These piles are joining by bolts or by means of dowel wedges. A bolted joint is capable to take up bending moments, shearing and pulling forces, and dowel-wedged joint is only capable to take compressive force.

The lower element of the piles has a symmetrical tip to provide the vertical position of the penetration of piles to be jacked. Sometimes the tip is reinforced with a tetrahedral symmetrical drive point of steel sheet 6 to 10 mm thick. All pile elements are reinforced with main longitudinal reinforcement provided with transverse collar.

*Table 3.3*

**Range of multisection piles designed by NIIpromstroy**

| Identification mark of pile section | Longitudinal reinforcement | Material consumption per section | | Weight of one section, kg |
|---|---|---|---|---|
| | | concrete, m³ | reinforcement, kg | |
| *With bolted joint* | | | | |
| CM 0,6.30-б | 4Ø16 A300 | 0.054 | 16.1 | 151 |
| CM 0,9.30-б | 4Ø16 A300 | 0.081 | 18.3 | 221 |
| CM 1,2.30-б | 4Ø16 A300 | 0.108 | 20.5 | 290 |
| *With dowel-wedged joint* | | | | |
| CM 0,6.30-ш | 4Ø16 A240 | 0.054 | 7.41 | 142 |
| CM 0,9.30-ш | 4Ø16 A240 | 0.081 | 8.42 | 211 |
| CM 1,2.30-ш | 4Ø16 A240 | 0.108 | 9.43 | 280 |

The lowered piles are joined with the adjacent bearing structures of a building as follows (Figure 3.9). Under the old foundation, a distributing element is installed that distributes uniformly a load from a jack during the pile penetration (pile reaction), due to which the lower surface of the wall constructions or strip foundations adjoins uniformly to the surface of the element even after the additional loading of the pile. The distributing element serves, at the same time, as a support of the hydraulic jack during the pile penetration. Between the distributing element and the pile head, a head element is placed that serves as a mount for the hydraulic jack during the operations. The enlarged pile head is completed with previously fabricated struts that are installed after the penetration.

In order to provide the vertical position of the Mega piles, the distributing elements shall be horizontal and tightly adjoined to the old foundation. For this reason, the surface of the old foundation is first leveled, and then a leveling coat of cement grout of the required thickness is placed between the old foundation and the distributing element.

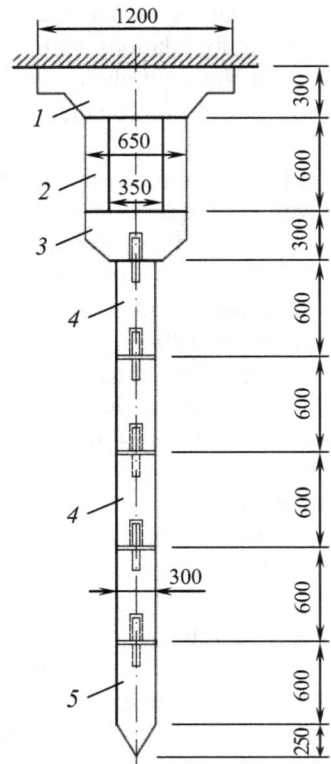

Figure 3.9. Mega pile design:
*1* – distributing element; *2* – strut;
*3* – head element; *4* – line element;
*5* – lower element

When underpinning strip foundations with the Mega piles (Figure 3.10, *a*), the works are started from the excavation of a pit and the sheeting of its walls. The pit depth depends on the selected length of the elements. The pit bottom shall be at least 1.5 m lower than the base of the old foundation. The pit depth depends also on the groundwater table.

The distributing element is arranged on supports under the old foundation base, and a leveling coat of fast-setting cement grout is placed between the supports.

On the pit bottom, the lower element of the pile is installed in the strictly vertical position. After that, the head element is placed on it to install the hydraulic jack. Instead of a heavy head element, it is appropri-

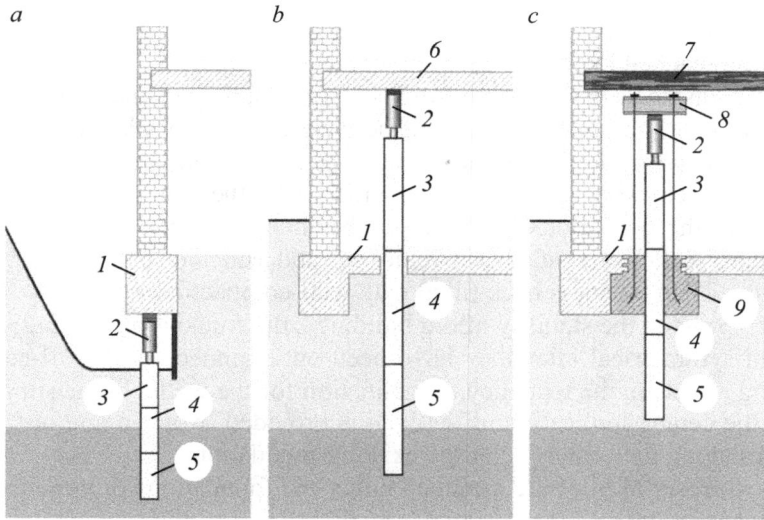

Figure 3.10. Options of underpinning with Mega piles:
*a* – due to thrust of jack against foundation to be underpinned; *b* – due to thrust of jack against basement flooring structure; *c* – due to anchorage of jacking mechanism in foundation to be underpinned; *2* – jack; *3* – head element;
*4* – line element; *5* – lower element; *6* – reinforced concrete flooring structure;
*7* – wooden flooring structure; *8* – thrust beam; *9* – cast-in-situ raft

ate to use a light but sufficiently stiff steel pad. The hydraulic jack is arranged between the head and distributing elements. A gap between the hydraulic jack and distributing element shall be filled with steel plates. After the vertical position of the jack and element has been checked, a hydraulic pump is switched on that drives the jack.

Under the load, the lower element with the point drives gradually into the ground. It will not be driven completely until the jack is relocated three or four times, as the stroke of the jack piston is 170 to 200 mm, and the driven element is 600 mm long. When the piston stroke runs short, the jack is relocated onto a new pad approximately 200 mm high and the pile driving is continued.

After the pile element with the point has been driven, the jack and pads are removed and the next line element is installed. A leveling coat of fast-setting cement grout is placed between the contacting end surfaces of the elements.

The joining and driving of the elements is continued until the design bearing capacity of the pile reaches the specified value of the driving load, which is determined by the readings of a pressure gauge installed on the jack. Upon reaching the specified load, the driving is stopped, and the preparation is started to install the head element of the pile. For this purpose, the reached load shall be fixed by means of struts.

If a distance between the head and distributing elements is longer than the strut length, the strut shall be wedged by steel plates of the required thickness, and if it is shorter than the strut length, the distance is to be increased by way of further driving the pile to the required depth.

After the head element of the pile has been completely wedged, the hydraulic jack is unloaded and removed, and concrete is placed at that location. Then the pit is back filled with well-compactible soil.

To provide the stability of old buildings, the foundation loading shall remain symmetrical after they have been underpinned, as the off-center loading results in the tendency to inclination for the walls. In the simplest case, the center application of the load is provided by arranging one row of piles along the centerline of the existing foundation.

Two rows of piles are arranged under the foundations of large thickness (at least 0.9 to 1.0 m), symmetrically relative to the foundation centerline in cases when the load from the old buildings is high and it is not sufficient to install one row of piles (Figure 3.11).

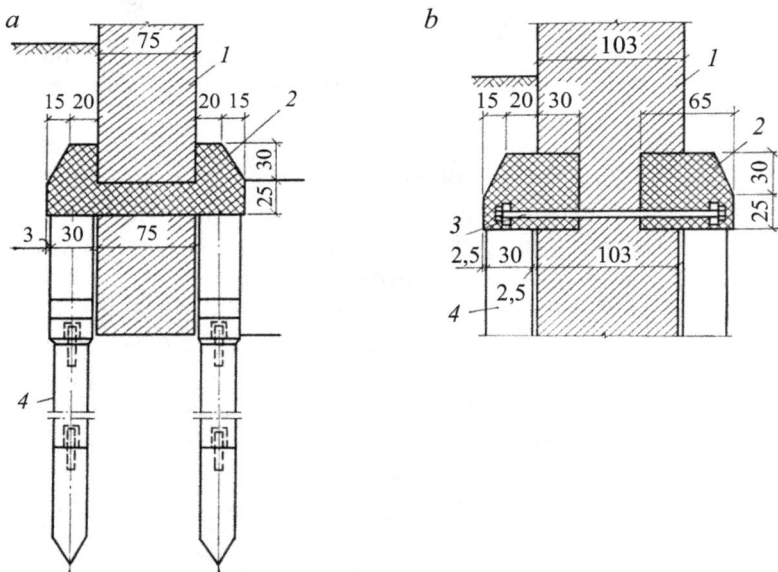

Figure 3.11. Options of underpinning strip foundation
with two rows of Mega piles:
a – with one beam; b – with two beams; 1 – existing foundation;
2 – reinforced concrete beams; 3 – tie rod; 4 – jacking pile

With a close distance between the pile rows, the first row that is driven at the beginning densifies soil to such extent that the driving of the second pile row becomes difficult. In view of that, the pile rows should be moved apart to a longer distance. On both sides of the foundation, reinforced concrete beams are placed that project from the wall in cantilever fashion. Sections 1.5 to 2.5 m long are prepared for the beams.

The pile rows are arranged under the cantilevers, and eventually the reaction from the piles are transferred to the reinforced concrete beams in an off-center way. To take up the forces occurring due to the off-center application of the load, the beams are connected with steel tie rods. A force in the tie rods is determined on the condition that the torque moment and the moment of stability are equal to each other. The tie rods can be made of common reinforcement steel, and they are anchored by means of a length of channel bar and a nut. The tie rods are installed prior to concreting the beams, in indents slotted in the foundation, and the anchor arrangements are placed near the external side surfaces of the beams.

The beams serve as supports for the jack at the time of the pile driving. Wherever possible, it is required to drive simultaneously two opposite piles. In case of a thick foundation and a high load on it, it is allowed driving piles separately.

To install two rows of piles instead of continuous longitudinal beams, local enlargements might be only made under the piles. It is very difficult to provide such enlargements when concreting in situ, therefore it is practical to use prefabricated enlargement structures. In this case, it is critical to provide the tight adjoining of the enlargement element to the upper surface of the indent made in the foundation. It can be done when the wall foundation is sufficiently strong and can be loaded with concentrated loads that are transferred via relatively small areas. Thus, it is preferred to install the continuous longitudinal beams, that distribute the reaction from the pile to a long distance and contribute to an increase in the longitudinal stiffness of the wall.

If there is a reinforced concrete slab available, the Mega piles are driven in the basement through holes with the dimensions of 30×30 cm slotted in the foundation slab, with cutting the main reinforcement of the slab (Figure 3.10, *b*). The basement flooring structure can serve as a thrust for the jack.

The advantages of this method lies in the fact that the scope of the works related to slotting the holes in the reinforced concrete foundation slab is small; however, this work is labor-consuming, expensive and takes considerable time.

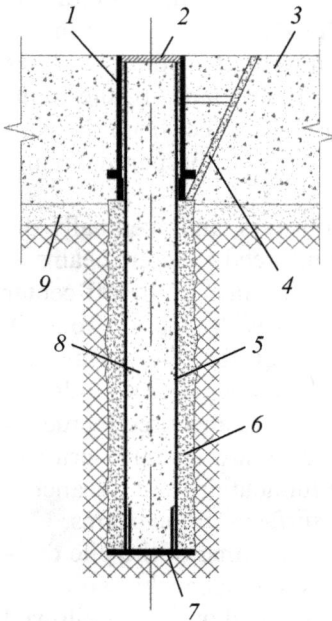

Figure 3.12. Soles pile design:
*1* – conductor; *2* – pile head
slab; *3* – reinforced concrete
slab; *4* – injection tube;
*5* – metal pipe; *6* – shell of injected cement grout; *7* – pile
shoe; *8* – concrete shaft of pile;
*9* – bed

The principal disadvantage of this method is the partial or complete termination of the use of the basement during operations. At the high groundwater table, there is a danger of waterproofing failure and water penetration into the basement. As it is not possible to completely fill up a gap around a pile and to arrange hydraulic lock, water will be always, to one extent or another, entering the basement, which shall be taken into consideration when using the basement space.

When it is not possible to thrust the jack against the basement flooring structure on the grounds of strengths (see Figure 3.10, *c*), it should be used a jacking mechanism, which is anchored in a cast-in-situ reinforced concrete raft constructed under the foundation slab. At the first stage, holes with the dimensions of 1.0×1.0 m are made in the foundation slab. Then soil is excavated from underneath the slab foot and the raft concreting is started, with providing a hole in the middle of the raft to let piles pass. The piles are driven using the jacking mechanism consisting of the jack, metal beam and metal rods. The metal rods welded to starter bars of the raft are tied to the beam, which serves as thrust for the jack. The disadvantages of this method are the extreme complexity, high cost and duration of operations.

**Soles piles.** The construction technology for these piles has been developed in Italy. The design of a Soles pile is a cast-in-situ concrete shaft in a metal pipe and a shell of injected cement grout (Figure 3.12).

The Soles pile construction technology implies the construction of a cast-in-situ reinforced concrete raft provided with pockets (holes) to make piles. To have the pile foundation under the reconstructed building to be involved in the work, the raft slab is brought under the foundation slab.

The raft construction begins with its reinforcing. After the reinforcing, pockets are provided in the raft slab (Figure 3.13). For this purpose, a metal shoe is placed at the level of the lower band of the raft between rebars, and a guide conductor made of a piece of metal pipe 146.0 to 355.6 mm in diameter is freely installed on the metal shoe. The conductor, is height shall exceed 10 to 30 cm, the thickness of the foundation slab. To prevent the conductor from tripping at the time of concreting, the conductor is tied to the reinforcing cage of the raft. To the external surface of the conductor, a collar rim is welded that prevents vertical displacements of the conductor when a pile is being driven. Prior to concreting, the conductors are covered with plastic caps. Around the conductors, embedded

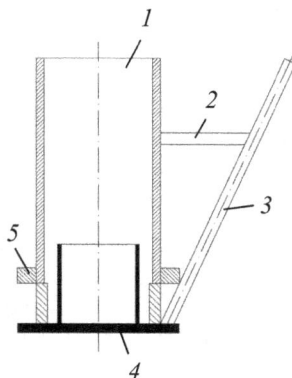

Figure 3.13. Pocket design:
1 – metal conductor; 2 – bar to fasten injection tube; 3 – injection tube; 4 – pile shoe; 5 – collar rim

parts for the anchorage of the jacking mechanism are fastened to the reinforcing cages. Then the raft slab is concreted.

The pile construction is started after the raft concrete has reached 80 % of the design strength. Over the pile construction location, the hydraulic jacking mechanism is installed and anchored to the embedded parts of the raft. After that, the first pipe section is installed in the conductor, and the first section together with the shoe is started to be jacked. At the time of jacking, the raft slab serves as weight for the hydraulic mechanism. After the first pipe section has been driven, its upper end is jointed by welding to the lower end of the second section and the jacking is continued. The jacking with the consecutive jointing of the metal pipe sections is carried out down to the required depth.

As the diameter of the flat shoe is greater than the diameter of the metal pipe, a cavity is formed around it, and cement grout is injected under pressure through injected tube into the cavity. The formed cement shell protects the metal pipe against corrosion and increases the friction of the pipe on soil. The internal cavity of the metal pipe is filled up with fine-aggregate concrete mix that forms the concrete shaft of the pile.

Depending on the stress conditions and the required bearing capacity of the pile both in terms of soil and material, there are used metal pipes 114.3 to 298.5 mm in diameter with the walls 8.0 to 12.5 mm thick. Flat drive points are produced 200 to 700 mm in diameter and 20 to 60 mm thick.

# Chapter 4. QUALITY CONTROL OF PILING WORKS

The quality control of construction products is an integral part of the construction operations. The high quality and the adequate cost of construction products are considered as principal factors that form the positive image of a construction company.

The piling works are controlled by means of the incoming, operating and acceptance inspection. The inspection is carried out by a construction supervisor, designing company (designer's supervision) representative and the Client's representative. If required, the Client shall engage a specialized scientific-and-research agency.

Particular care shall be paid to the quality control related to the hole-making and pile-concreting. In poor-quality construction of bored and cast-in-place displacement piles, the major danger is the reduction in their bearing capacity and, as a consequence, the development of deformation and loss of stability for a building.

The bearing capacity of end-bearing piles depends on the strength of pile material and soil under the pile foot. Therefore when constructing the end-bearing piles, it is required to pay attention to the compliance of the properties of actual materials used to concrete piles with the rated ones (incoming inspection). Besides, it is required to eliminate the possibility of the narrowing formation due to borehole wall creeping, the breaks in concreting and the concrete mix disintegration and to monitor the leak tightness of the casing pipe connections when concreting pipes in water-saturated soils (operating control).

For floating piles, the most hazardous is the reduction in the pile bearing capacity due to the non-penetration of a borehole through the operators' fault and, as a consequence, the reduction in the design length of the pile. Another cause for the reduction in the pile bearing capacity might be the poor-quality removal of drilling mud returns from the borehole bottom. The above defects can be avoided by means of the operating control of piling works.

## 4.1. Incoming inspection

The incoming inspection is an inspection of incoming materials, products, structural units, etc. as well as technical documentation.

For concrete mix, reinforcement, reinforcing cages, casing and tremie pipes, reinforced concrete slabs for temporary roads, soil and other materials and products, the incoming inspection is performed by way of their visual examination. During the examination, received materials, products and structural units are checked for their compliance with the regulatory and design requirements. It should be also checked the availability and content of covering documents (technical certificates, certificates, bills of lading). All details of the received materials, products and structural units are recorded in an appropriate logbook. The covering documents are attached to the logbook.

When constructing bored and cast-in-place displacement piles, there shall be controlled the placeability (consistency) of concrete mix and the concrete strength. When carrying out works in winter conditions (at temperatures below freezing point), the concrete-mix temperature shall be controlled taking into consideration the recommendations given in SNiP 3.03.01. *The fabrication quality of a reinforcing cage* shall meet the requirements of a design and GOST 14098.

*The consistency of concrete mix* shall be controlled by the cone slump in accordance with GOST 10181. For this purpose, samples of concrete mix shall be taken when placing the first batch of concrete mix into a borehole and upon completion of placing every 5 m$^3$. Samples shall be taken more frequently if the consistency of concrete mix is visually different from the required consistency. If the difference is greater than 2 cm between the actual and design consistency of concrete mix, it is not allowed placing concrete mix.

*The concrete strength* shall be determined at the design age by the destructive method in accordance with GOST 18105. For this purpose, at least two samples of every concrete batch and no less than one sample a day shall be taken out of randomly selected batches, in accordance with GOST 10181, when concreting piles. Out of every sample, one series of each concrete samples are made, in accordance with GOST 10180, in calibrated moulds that meet the requirements of GOST 22685.

Based on the test results for the reference samples, a report shall be made that shows the actual concrete mix. In accordance with GOST 18105 and SP 50-102, the reference samples of pile concrete shall harden in the same conditions as the conditions of concrete hardening in borehole. To this effect, the reference samples shall be stored on a pallet in a specifically drilled borehole. The position depth of the pallet with the reference samples shall approximately correspond to the position depth of the concrete mix batch, from which the samples were taken. As a matter of fact, the reference samples harden outdoors in the conditions that are different from the conditions of concrete hardening in boreholes. Even if the reference samples harden on the pallet in the borehole, this method cannot be considered reliable, as the technology of placing concrete mix into borehole, its compacting and the conditions of the concrete hardening in the pile body considerably differ from the method of making reference cubes. Therefore, the most representative are the strength test results for core sample drilled out of the pile shaft.

As agreed with the design company that carries out the designer's supervision, it is allowed not taking the concrete mix samples at the location of their placement in the borehole, but determining the concrete strength based on the control data provided by the concrete mix manufacturer.

## 4.2. Operating control

**Pile penetration.** The pile penetration, which is understood as driving a finished pile or constructing a pile directly in the ground, is started after the instrumental check of the surface level and the pile position on site.

During the pile penetration, it shall be controlled a departure of the pile longitudinal axis from its design position.

When drilling and reaming the boreholes, the following shall be recorded: borehole penetration rate, vertical position and rotation speed of a working tool, as well as forces applied to it (pressing force and torque). If required, the soil survey is to be performed when drilling boreholes. For this purpose, the agency shall be employed, which carried out the engineering and geological survey on the site.

When making a borehole by way of static displacement, the following shall be controlled: clamping and jacking forces acting on a shell (pipes with closed bottom end), its vertical deviation, penetration rate and depth.

At the time of the driving or vibrational penetration of a shell, it shall be recorded its deviation and refusal. A refusal shall be understood as a value of the shell settlement resulted from one hammer blow when driving the shell and as a value of the shell settlement resulted from the operation of vibratory pile driver for 1 min when operating the vibratory pile driver. The recommendations given in SNiP 2.02.03 or in SP 50-102 are to be used to estimate the bearing capacity of a pile based on the data recorded in the driving or vibrational penetration of the shell.

All data recorded when driving or constructing piles shall be recorded in the required logbook, the entries of which shall be controlled by the designing company (designer's supervision) representative and/or by the Client's technical supervision representative.

Modern rigs are equipped with on-board computers (Figure 4.1) that allow continuously monitoring the hole-making parameters (Figure 4.2) and saving them in memory. A file is assigned with a name that corresponds to the pile number shown in the as-built diagram of a pile field. Some rigs are provided with printers that print a diagram chart with the parameters recorded at the time of hole-making.

Figure 4.1. On-board computer in cab of Bauer drilling rig

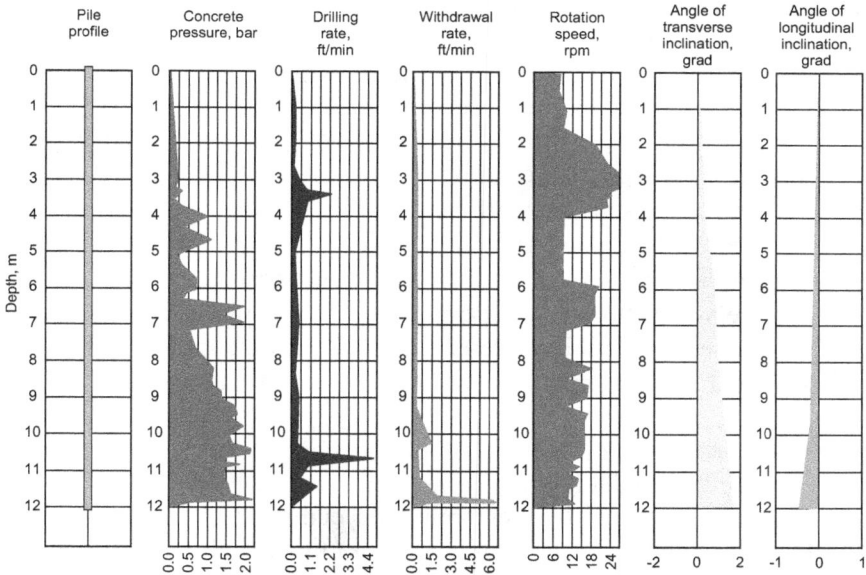

Figure 4.2. Data on progress of pile constructing displayed on screen
of on-board computer

Upon completion of making a borehole, its depth and the quality of removal of drilling returns from the borehole bottom shall be controlled by way of slowly lowering a working tool of the drilling rig and taking samples from the borehole bottom. The quality of the borehole cleaning can be examined by means of a small-size digital video camera. A deviation of the borehole depth from the design value shall not exceed ±100 mm. A report shall be made based on the results of the borehole survey.

**Reinforcing and concreting.** The reinforcing and concreting shall be started if the required report on the borehole availability is in place.

During the pile reinforcing and concreting operations, a construction supervisor shall maintain the required logbook, the entries of which shall be checked by the designer's supervision representative.

Prior to reinforcing a pile, there shall be checked the compliance of the reinforcement or reinforcing cage with the design requirements. The reinforcing cages that are factory made or directly constructed on the construction site shall be provided with the marking in addition to a technical certificate. In the logbook, there shall be recorded the number of the reinforcing cage to be installed in the borehole.

For the piles that are constructed without lost tip or shoe and reinforced throughout their length, provision shall be made to prevent the structure of the soil at the borehole bottom to be disturbed when installing reinforcement bars or reinforcing cage.

For the purpose of reinforcing the upper part of a pile, bracing shall be provided that prevents the reinforcement from moving when extracting the casing pipes and compacting concrete mix.

The following shall be continuously monitored when concreting the pile:

- consistency of concrete mix;
- concrete-mix delivery pressure;
- concrete-mix placement rate;
- concrete-mix level in tremie pipe and in borehole;
- levels of bottom ends of tremie and casing pipes (bottom end of tremie pipe shall be lowered into concrete mix at least by 1 m);
- concrete-mix temperature (at the outdoor temperature below zero);
- conformance of volume of placed concrete mix to volume of concrete-mix`column in casing pipe.

The time of concreting beginning and end shall be recorded in the construction logbook. In the same logbook, there shall be recorded the forced interruptions in concreting, their causes and duration.

When constructing bored injection piles, there shall be monitored the pressure and duration of the borehole pressure testing.

When concrete mix is electrically cured, its temperature shall be measured at the pile head using industrial thermometers embedded in concrete mix.

During the first four hours after the beginning of the electrical curing, the temperature should be measured at intervals of an hour, and during the period of the isothermal heating and cooling, it should be done in accordance with the guidelines given in SNiP 3.03.01.

The hardening conditions of concrete mix shall be controlled during the heating period. The results of the temperature measurements shall be recorded in the logbook.

Bored and cast-in-place displacement piles concreted in removable shells are constructed in a sequential order, one after another, with a distance between their longitudinal axes being longer than $6d$ (where $d$ = maximum diameter of pile cross section). If the distance between the longitudinal axes of the piles is equal to or shorter than $6d$, then the construction of the following pile shall not be started until the concrete of the adjacent piles reaches a strength of 2.5 MPa. To compute the concrete strength $R_t$, MPa, at a certain age, the following formula shall be used

$$R_t = R_{28} \frac{\lg t}{\lg 28},\qquad (4.1)$$

where $R_{28}$ = concrete strength at the age of 28 days, MPa; $t$ = age of concrete in days, at least 3 days.

## 4.3. Acceptance inspection

### General provisions

The pile foundation structures shall be accepted on the basis of the results of an acceptance inspection based on the design, as-built and shop documentation. The acceptance inspection is required to verify the compliance of the completed foundation structures with the design and the requirements of regulatory documents.

Piles and pile foundations are controlled and accepted by the Client's Technical Supervision Department with the participation of the pile foundation designers and contractors who performed the foundation construction works.

The acceptance of the pile foundations shall be carried out in two stages: after the piles have been driven or constructed and after the construction works have been completed for pile caps. At that, the construction of pile caps shall not be started until the pile field is accepted.

The acceptance of the works related to the construction of the foundation structures shall be done based on the following:

- designs of pile foundation and method statements;
- process procedures for works;
- manufacturers' technical certificates for driven piles and prefabricated pile caps, reinforcement and concrete mix (ready-mixed concrete) for piles and cast-in-situ pile caps constructed on site;
- incoming quality control logbook for materials and structures;

- general-purpose work logbook;
- report on handover/acceptance of foundation pit for driving or construing piles;
- report on surveying of gridlines for building and foundations and setting pegs on construction gridlines;
- laboratory test reports for reference samples of concrete;
- as-built diagrams of pile layout showing deviations of piles in plan view, in depth and in vertical position;
- pile driving or construction logbooks;
- summary sheets for driven or constructed piles;
- documentation based on results of pilot surveys including results of soil tests with piles in accordance with GOST 5686;
- reports on survey of reinforcing cages and boreholes prior to concreting of piles constructed on site.

At the time of accepting the works related to the construction of the pile foundation structures, it is required:

- to study submitted documentation;
- to carry out the survey of piles and to verify the compliance of the completed works with the design;
- to check, using instruments, the correct position of the piles;
- to carry out the check tests of piles if their bearing capacity raise doubts.

According to the Russian regulations (SP 50-102), it is required, when accepting the piles, to check their position in plan view, the pile head levels and the vertical position of the pile centerlines.

In case of the one-row arrangement of driven, vibratory-driven, jacked and screw piles, the limit deviation of the actual position of the piles in plan view from the design position is $\pm0.2d$ ($d$ = diameter or side of pile cross section) across the centerline of the pile row and $\pm0.3d$ along the centerline of the row; in case of clusters and strips with two- or three-row arrangement, $\pm0.2d$ for border piles across the centerline of the pile row and $\pm0.3d$ for the other piles and the border piles along the centerline of the pile row; in case of a compact pile field, $\pm0.2d$ for border piles and $\pm0.4d$ for middle piles.

In case of cast-in-place displacement, bored and bored injected piles, the limit deviation of the actual position of the piles in plan view from the design position is $\pm10$ cm across the pile row and $\pm15$ cm along the pile row in case of the cluster arrangement of the piles.

In case of a cast-in-situ raft or slab, the limit deviation of the actual levels of pile heads from the design levels is $\pm3$ cm; for a prefabricated raft, it is $\pm1$ cm, and for a foundation without raft with prefabricated pile cap, it is $\pm5$ cm.

For driven or constructed piles, the limit deviation of their center-lines from the vertical position is ±2% of their length.

In addition, the quality of shafts shall be controlled for cast-in-place displacement, bored and bored injected piles by means of destructive and non-destructive inspection methods.

To check the possibility of transferring to piles the loads specified in the design, there shall be carried out the field check tests of the piles under static load.

Based on the results of the acceptance inspection of piling works, there shall be made a report where all identified defects shall be recorded and the corrective action shall be recommended.

After an additional computational justification, the pile foundation designer can determine whether it is possible to use piles with defects or deviations beyond the limit values.

The construction of pile caps shall not be started until the pile field is accepted.

## Destructive methods of pile quality control

*The shaft quality control for bored and cast-in-place displacement piles* lies in strength-testing of core samples vertically drilled out at different depth at an interval of 0.5 m. For this purpose, piles shall be selected in accordance with GOST 28570. In particular, the piles shall be selected for taking concrete samples after a visual examination depending on the stress condition with allowance for the minimum possible reduction in their bearing capacity.

The samples are taken using a small-scale rotary drilling rig, which is installed over the pile and anchored in the soil base.

For core sampling, a drill bit set with diamonds and hard alloys is screwed on the end of a core tube. Hollow drilling rods are held by a drill chuck fastened to the end of a hollow threaded core tube. To cool down the drill bit and to remove crushed stuff and drilling returns, drilling mud is delivered downward via the hollow drilling rods. A concrete cylinder of undisturbed structure (core) is formed, as the drilling bit advances in the core receiver of the core tube. When the drilling bit reaches the required depth, the stem, core tube and drilling bit are extracted from the borehole and the core is removed from the core tube. A cavity formed in the borehole shaft shall be filled with fine-aggregate concrete.

After that, the cores shall be marked (Figure 4.3), and there shall be made a sampling report. The report shall include the number of the pile, from which the sample was taken, the absolute level of the pile head at the time of sampling, the quantity of the samples taken and their diameter, depth and sampling date.

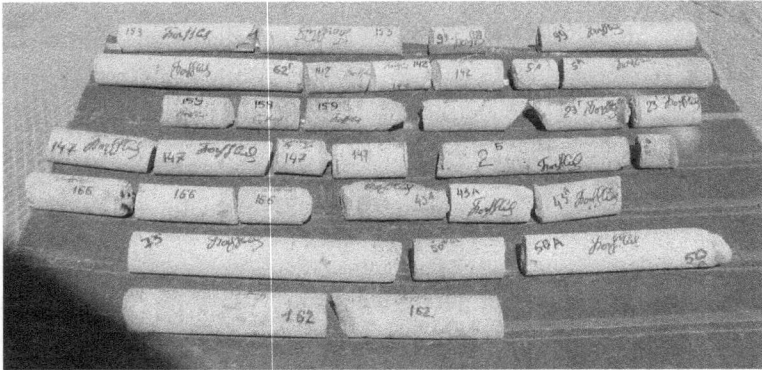

Figure 4.3. Marked cores drilled out of pile shafts

Then the cores are sent to a workshop where they are sawed by core specimens of the required height in accordance with GOST 10180. The end faces of the core specimens are polished, following which the specimens are sent to a laboratory for compression test. The shaft quality control shall be performed for one pile out of hundred but not fewer than two piles out of the total quantity.

*For bored and cast-in-place displacement piles, the shaft integrity check* can be done by way of estimating the specific water absorption of concrete. For this purpose, a hole is drilled inside a pile along its longitudinal axis, and the hole is water-pressure tested. Usually, there are used holes, from which cores were taken.

Prior to water-pressure testing, the holes shall be thoroughly cleaned flushed out to remove drilling returns. At the beginning, the water-pressure testing shall be carried out for the entire hole. The specific water absorption (l/min) is calculated per 1 m of the hole at 1 atmosphere of the pressure by the following formula

$$\gamma = \frac{Q}{\left(H + \dfrac{h}{10}\right)lt},$$ (4.2)

where $\gamma$ = specific water absorption of concrete, l/min; $Q$ = total water flow rate during water-pressure testing, l; $H$ = pressure gauge readings, atm.; $h$ = water column height from pressure gauge centerline to middle of hole section to be tested, m; $l$ = depth of hole section to be tested, m; $t$ = time after stabilization of water flow rate, during which pressure is maintained at specified level (at least, for 30 min), min.

If the water absorption does not exceed 0.1 l/min, the concrete integrity is considered satisfactory. Otherwise, the water-pressure testing shall be carried out on individual sections of the hole to identify defect locations. Depending on the depth of a defect location and its approximate dimensions, there shall be made a decision on injecting cement grout into the hole or making a redundant pile.

*The shaft quality control for bored injected piles* is carried out by way of digging up the heads of 2 % of the completed files and testing them for strength by non-destructive methods (GOST 17624, GOST 22690). When detecting defects in the piles to be tested, the number of tests shall be increased.

In case of bored and cast-in-place displacement piles, it is not possible, in practice, to implement the shaft quality control for all piles by way of testing concrete cores for strength as it is destructive, very time-consuming and expensive method.

Meanwhile the piles – critical structures, on which quality depends the operational reliability of a building (facility) – shall be subject to one-hundred-percent inspection as it is done at concrete products plants.

## Non-destructive methods of pile quality control

For the purposes of one-hundred-percent inspection, non-destructive methods shall be used. Seismic-acoustic (sonic) and ultrasonic inspection methods are used to determine the actual length of piles, to localize defects (cracks, "necks" – weakening of cross sections) and to evaluate the mechanical properties of pile concrete.

Figure 4.4. Testing of bored pile with IDS-1 seismic-acoustic instrument

The work with seismic-acoustic and ultrasonic instruments is divided into two stages: testing of piles on construction site (Figure 4.4) and interpretation of obtained data using dedicated software.

**Seismic-acoustic method.** The principle of operation of seismic-acoustic instruments is based on recording the parameters of elastic waves generated in piles by means of impact impulse transferred the end face of the pile (Figure 4.5).

After a blow of the hammer on the end face of the pile, a longitudinal tensile-compression wave propagates along the pile shaft at a certain velocity $c$. As the acoustic properties of concrete differ significantly from the acoustic properties of dispersed soils, the pile represents a wave duct with the relatively small energy loss by wave attenuation and by re-emission to geological body. The acoustic wave is reflected at the interface of media (concrete – foreign particulate, concrete – soil, etc.). The time interval between the initial hammer blow and the reflection from the interface of media is measured by an instrument and equal to $t$ that is required for the wave to propagate twice along the pile shaft of length $l$ (downwards and upwards):

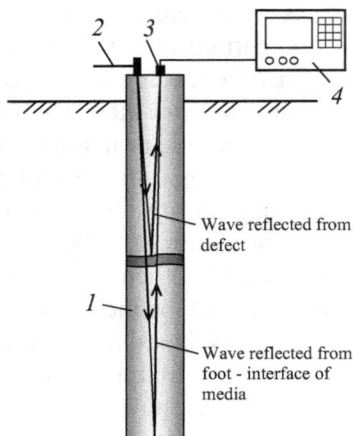

Figure 4.5. Principle of Seismic-acoustic method: 1 – pile; 2 – hammer; 3 – seismic receiver; 4 – instrument

$$t = \frac{2l}{c}. \qquad (4.3)$$

After the time $t$ of the acoustic wave propagation along the pile has been measured, one of two parameters shall be determined:

- wave propagation velocity along pile length;
- pile length by the known wave propagation velocity.

To record reflected waves or return signals, there shall be used sensors of electrodynamic or piezoelectric types installed on the end surface of the pile. To improve the acoustic contact between the pile and seismic receiver, there shall be used special-purpose non-freezing mastic compounds or plasticine.

Being generated by an exciter and recorded by a velocimeter (velocity transducer), a signal is transformed by means of analog/digital converter into a reflectogram – variation of the pile head particle displacement velocity $V$ with time. If an accelerometer (acceleration transducer) is used, then the instrument automatically integrates the acceleration-time relationship to obtain a velocity-time relationship.

The reflectogram are stored in the form of files in the instrument memory. After the pile testing, the files are transferred to a personal computer. Then using the software supplied together with the instrument, the reflectograms are processed – the signal is filtered and amplified. The filtration allows removing high-frequency noise that makes it difficult to make analysis of the testing results and obtaining a smooth reflectogram.

The friction along the side surface of the pile causes the signal attenuation along the length. In order to obtain a clear reflectogram, the friction that is generally considered as logarithmically increasing with depth is compensated using the time-amplitude adjustment of the amplification. Typically, the exponential amplification is used for this purpose. In well-balanced reflectograms, the peaks, which correspond an impact on the pile head and the reflection of a wave from the pile foot, shall have an approximately equal amplitudes (Figure 4.6). The analysis of the reflectograms allows determining the pile length and localizing defects in its shaft.

Pile No.: 15

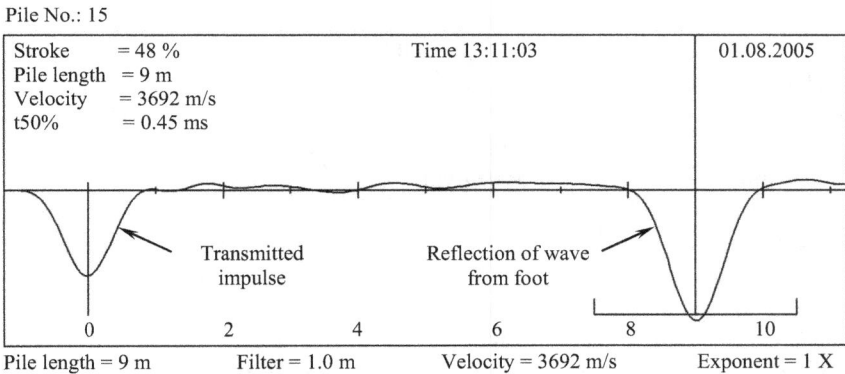

| Stroke = 48 % | Time 13:11:03 | 01.08.2005 |
| Pile length = 9 m | | |
| Velocity = 3692 m/s | | |
| t50% = 0.45 ms | | |

Transmitted impulse        Reflection of wave from foot

0     2     4     6     8     10

Pile length = 9 m        Filter = 1.0 m        Velocity = 3692 m/s        Exponent = 1 X

Figure 4.6. Reflectogram obtained during tests of driven pile using
Integrity Testing System seismic-acoustic instrument (pile length = 9 m;
acoustic wave velocity = 3692 m/s; oscillation period = 0.45 ms)

The pile length $l$ is determined by an indirect method on a basis of the time interval $t$ measured with instrument; at that, the longitudinal wave velocity $c$ in the pile is considered known:

$$l = \frac{ct}{2}. \qquad (4.4)$$

The error of determining (that is to say, determining and not measuring) the pile length $l$ (wave velocity $c$) directly depends on how accurately the wave velocity $c$ (pile length $l$) is specified.

In order to obtain consistent results, the end faces of the piles shall be horizontal, clean with a roughness of no more than 2.0 mm. No water, cement grout and cracks are allowed on the end faces of the piles. The concrete age shall be at least 7 days at the time of testing. During the pile testing period using the seismic-acoustic method, it is not allowed operating machinery that generates vibration.

If a wave is assumed as an independent object, then, using the Newton's second law, it can be shown that the movement of an elastic wave along a bar is characterized with the sound velocity, which depends on the medium properties and is determined by the following formula

$$c = \sqrt{\frac{E}{\rho}},$$ (4.5)

where $E$ = dynamic modulus of concrete elasticity, MPa; $\rho$ = concrete density, kg/m$^3$.

In this case for heavy B35 concrete of natural aging, the propagation velocity of the longitudinal wave is

$$c = \sqrt{\frac{34.5 \cdot 10^9}{2500}} = 3700 \text{ m/s}.$$

For construction materials, in particular, for concrete, no direct functional relationship exists, but there is sufficiently rigid and close correlation (i.e., statistic relationship) between the acoustic velocity and the material strength: the acoustic velocity is higher in a stronger material. In abbreviated form, this relationship is called "correlation – velocity – strength". However, this relationship does not reflect a complex concept of the concrete strength. Thus, transverse tensile waves and radial waves occur in piles (bars) when the longitudinal waves propagate. Therefore, the propagation velocity of the longitudinal waves is lower than in unbounded medium.

*The pile defects* can be characterized by a change in the cross sectional area from $A_1$ to $A_2$ or by the material properties $E$ and $\rho$. When a wave encounters some discontinuity, it partly reflects back and partly moves forward (Figure 4.7).

Assumed that only a longitudinal wave occurs under the impact action, the analysis of the wave behaviour in piles of arbitrary shape uses the one-dimensional wave propagation theory, according to which the stress generated by the reflected wave will be as follows:

$$\sigma_{refl} = -\frac{Z_1 - Z_2}{Z_1 + Z_2} \sigma_{inc};$$ (4.6)

and the stress generated by the passed wave will be as follows:

$$\sigma_{trans} = \frac{2Z_2}{Z_1 + Z_2} \frac{A_1}{A_2} \sigma_{inc},$$ (4.7)

where $Z_i = A_i c_i \rho_i$ = acoustic resistance (impedance) of $i$-th medium; $\sigma_{inc}$ = stress generated by wave as a result of impact action.

Equations (4.6) and (4.7) allow simulating the wave behaviour and can be applied to bars of arbitrary form. For the purpose of the visual representation of the wave propagation, it is convenient to use a function between a coordinate of the longitudinal wave and the time: $x = f(t)$.

The relationship between stresses $\sigma$ generated by the impact action and the pile head particle displacement velocity $V$ is determined by the following formulas:

$$\sigma = E\frac{\Delta l}{l}; \qquad (4.8)$$

$$V = \frac{\Delta l}{\Delta t}, \qquad (4.9)$$

where $\Delta l$ = displacement of pile particles, m; $\Delta t$ = time, during which displacement $\Delta l$ occurs, s.

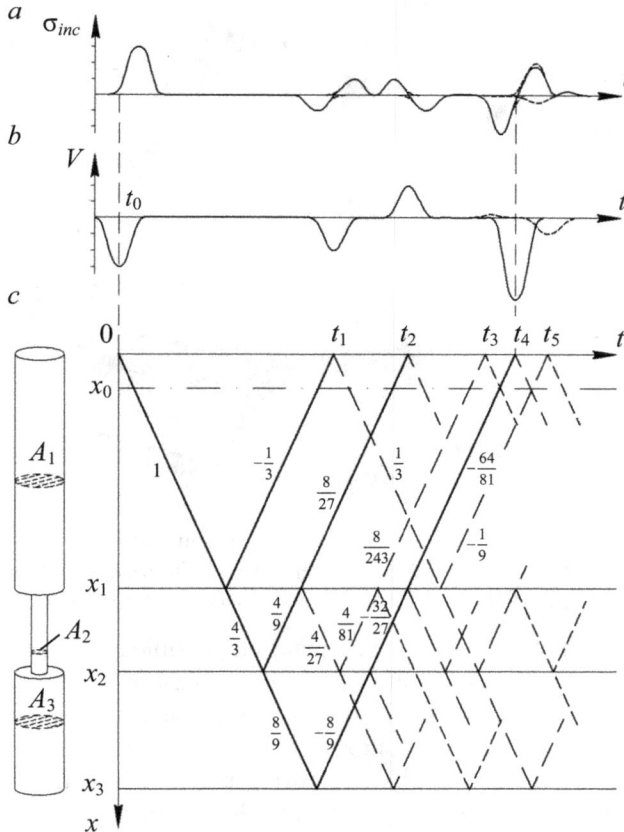

Figure 4.7. Pile with weakened cross section:
$a$ – variation of stress $\sigma$ in pile head with time $t$; $b$ – reflectogram (variation of pile head particle displacement velocity $V$ with time $t$); $c$ – variation of acoustic wave position with time (interference of waves)

103

Figure 4.7 shows the plot of the wave propagation in a pile with the weakened cross section. The stresses were calculated using Equations (4.6) and (4.7).

Various instruments (Figure 4.8) are used to test piles by the seismic-acoustic method.

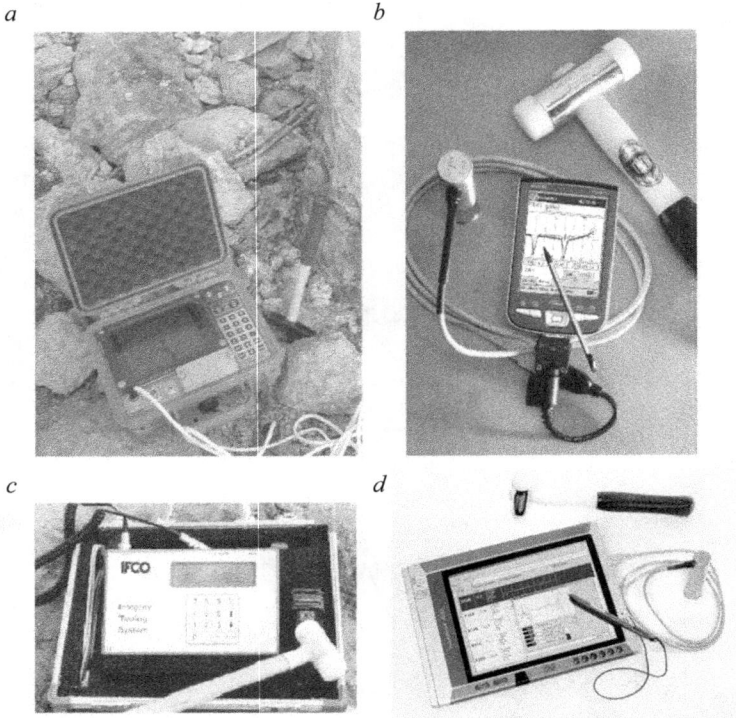

Figure 4.8. Seismic-acoustic instruments:
a – IDS-1 (Russia); Pocket Pile Echo Tester (Great Britain); c – Integrity Testing System (the Netherlands); d – Pile Echo Tester (Great Britain)

**Ultrasonic method** This method makes it possible to assess the integrity (continuity) of concrete shafts of bored and cast-in-place displacement piles. When applying the method, there are used sensors that emit the ultrasonic-range waves with the oscillation frequency higher than 20 kHz.

To perform the pile monitoring during their construction, plastic or metal tubes are embedded along the entire length of the piles. The tubes are arranged in parallel to each other. A diameter of every tube shall be such that the tube could accommodate a source or receiver of ultrasonic waves when testing a pile. A number of the tubes depends on the pile diameter. Figure 4.9 shows some options of the tube arrangement in a pile.

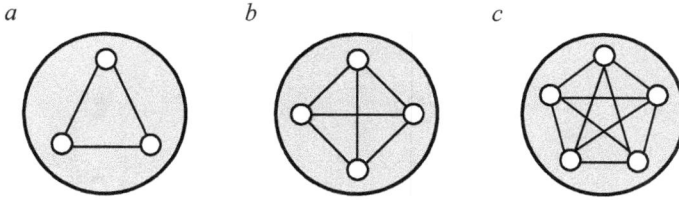

Figure 4.9. Tube arrangement in pile with ultrasonic inspection systems

The ultrasonic inspection is carried out using the instruments (Figure 4.10), which include sensors (transmitter and receiver of ultrasonic waves), a data recording device, a cable to connect the sensors to the recording device and attachments to take measurements of a sensor penetration depth. State-of-the-art instruments are provided with digital depth-gauges.

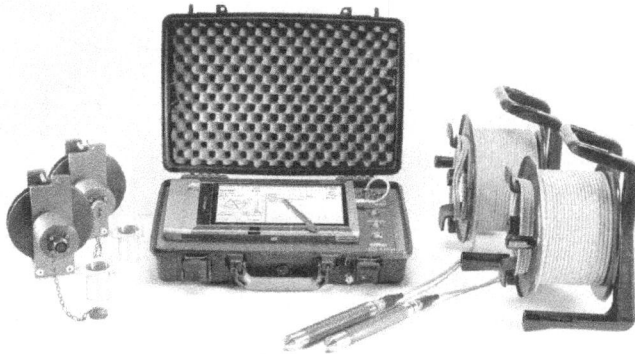

Figure 4.10. Cross Hole Ultrasonic Monitor

To generate waves, piezoelectric transducers are used that convert an electric impulse into acoustic oscillations. Precisely the same transducers convert an acoustic wave, which passed through the pile, into an electric signal. Prior to starting the tests, the tubes are filled with water, thus providing the acoustic contact between the transmitter and receiver of ultrasonic waves.

The ultrasonic inspection of pile-shaft concrete lies in the following. The ultrasonic wave transmitter is lowered to the bottom of one tube, and the receiver is lowered to the bottom of the other tube. After it has been generated, an ultrasonic wave propagates along the concrete shaft of the pile. The time of the first wave arrival is recorded by the receiver and saved in the memory of the recording device.

As the distance $l$ between the transmitter and receiver is known and the time $t$ of the ultrasonic wave progress is measured by the instrument, the ultrasonic propagation velocity in concrete between the tubes can be calculated by the following formula

$$c = \frac{l}{t}.$$ (4.10)

After the concrete has been inspected at a certain depth, the source and receiver are lifted up and the concrete inspection is repeated between the tubes. Concrete is consecutively inspected in such manner along the entire length of the pile. To reduce the testing time, multiple receivers of ultrasonic waves are used at the same time.

The instrument software allows visualizing the results of a series of measurements (Figure 4.11). Time-velocity curves of the ultrasonic propagation along the pile length allow assessing the integrity and the quality of a pile concrete shaft between two tubes.

To obtain a tree-dimensional tomogram of the ultrasonic propagation velocity, the transmitter and receiver are placed at different depths during the testing.

Upon completion of the ultrasonic testing, cement grout is injected into the tubes, by the same token increasing the bearing capacity of the pile concrete shaft.

It is appropriate to control the quality of the pile concrete shafts at the same age, at which the reference concrete

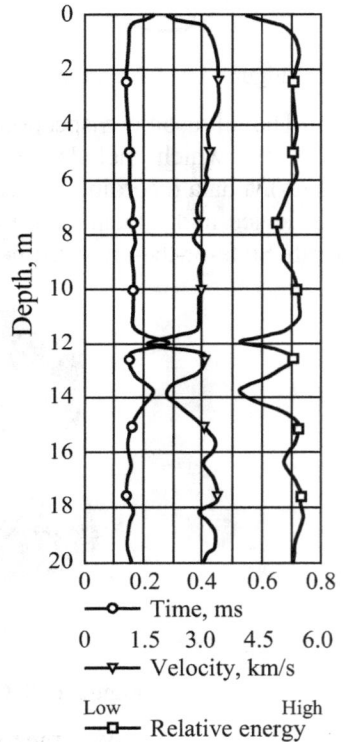

Figure 4.11. Ultrasonic inspection results for concrete between two tubes

samples (concrete cubes) and/or core samples were tested for strength.

For a certain concrete composition and technology of placing into borehole, the concrete strength can be estimated by rating curves. To plot the curves, the ultrasonic propagation velocity shall be determined in the core samples in accordance with GOST 17624 prior to durability testing the core samples for strength. By way of the mathematic treatment of the results of the mechanical tests and the ultrasonic inspection data, a velocity – strength correlation graph is plotted using the technique set forth in GOST 17624.

To approximately assess the concrete quality, the data are used that are given in Table 4.1 [4].

*Table 4.1*

**Concrete quality assessment by ultrasonic propagation velocity**

| Concrete quality | Ultrasonic propagation velocity, m/s | Compressive concrete strength, MPa |
|---|---|---|
| Very poor | < 3000 | 7 |
| Poor | 3000...3400 | 10 |
| Satisfactory | 3400...3900 | 18 |
| Good | 3900...4300 | 30 |
| Very good | > 4300 | > 30 |

**Ultrasonic propagation velocity in concrete and reinforced concrete** The ultrasonic propagation velocity in concrete can vary within a range of 2000 to 5500 m/s. Such variation is due to the effect of various factors. Among them, the most essential are the following: concrete density and age, hardening conditions, volume and type of concrete aggregate, water-to-cement ratio, moisture content of concrete, reinforcement percentage in reinforced concrete structural units, stress conditions of item to be controlled [4].

*Effect of concrete density* In case of constant concrete composition, its strength depends on porosity. The more pores are in concrete, the lower is its density and, therefore, its strength. Pores prevent the ultrasonic sound from propagating. For example, when increasing the concrete porosity by 10 %, the ultrasonic velocity reduces by 7 %, and when increasing the concrete porosity by 30 %, the ultrasonic velocity reduces by 30 %.

*Effect of concrete age* The concrete hardening, which is accompanied by the gain of its strength, generally occurs during the first month. The ultrasonic velocity increases as the concrete strength increases. Therefore the considerable increase in the ultrasonic velocity is only observed during the first month after concreting a pile or other structure.

Knowing the concrete age $t$, days, and its cube strength $R$, MPa, the ultrasonic propagation velocity c, m/s, can be calculated by the following formulas [42]:

$$c = c_{30} \left( \frac{R}{30} \right)^{1/6};$$
(4.11)

$$c_{30} = 3946 \lg^{1/6} (t+1),$$
(4.12)

where $c_{30}$ = wave propagation velocity, m/s, in concrete of compressive strength $R$ =30 MPa.

*Effect of amount and type of concrete aggregate.* Crushed rock or gravel of different mineral composition is used as coarse aggregate for concrete. Concrete is non-homogeneous material as the physical and mechanical properties of coarse and fine aggregates differ from the corresponding properties of cement stone. That determines the different ultrasonic propagation velocities in concretes of equal strength made with aggregates of different mineral composition. In view of the fact that coarse aggregate is 70 to 85 % of a heavy-concrete volume, the ultrasonic velocity in concrete depends, in a greater degree, on the elastic properties of aggregate at the constant values of the water-to-cement ratio and the amount of coarse aggregate.

In fine-aggregate concrete, the elastic properties of sand and cement have an effect on the ultrasonic velocity. Mineral composition of sand varies practically everywhere within insignificant limits, which affects slightly the ultrasonic propagation velocity.

The variation in sand amount have a considerable effect on the ultrasonic propagation velocity. Thus, an increase in the sand volume by 10 % causes a change of 5 to 10 % in the ultrasonic velocity.

*Effect of cement amount.* In concrete, cement serves as material that binds fine and coarse aggregates between each other. At that, the strength of cement stone is lower than that of aggregates. After exceeding a certain volume of cement that is required to bind aggregate, the concrete strength reduces. Thus, the ultrasonic propagation velocity is influenced by the cement amount, which, in its turn, has an effect on the concrete strength.

*Effect of temperature.* When temperature increases, the ultrasonic sound velocity decreases. For instance, when the concrete temperature increases by 10 °C, the ultrasonic sound velocity decreases by 40 m/s.

*Effect of reinforcement.* If there is reinforcement in concrete, the ultrasonic propagation velocity increases. In reinforced concrete, the ultrasonic velocity increases, in average, by 6 to 8 % as compared to unreinforced concrete depending on an amount and diameter of reinforcement used.

*Effect of concrete stress conditions.* The ultrasonic propagation velocity significantly depends on concrete stress conditions.

When compressing a concrete sample with the incremental load and concurrently scanning it with the ultrasonic sound crosswise relative to the action of a force or at small angle of inclination at the initial stage of loading, there can be observed an increase in the ultrasonic velocity. When the load increases further, the ultrasonic sound velocity begins to decrease. Such behaviour can be explained by the following circumstances. At the initial stage of loading, when stresses in concrete are low, concrete compaction occurs, and this causes an increase in the ultrasonic sound velocity. Under high stresses, cracks begin forming in concrete, and this process grows as the loads increase. As the cracks form and their number and length increase, the ultrasonic velocity decreases.

When stretching a concrete sample, no concrete compaction occurs, and therefore with the gradual load increase, the ultrasonic velocity continuously decreases.

Carrying out mechanical and ultrasonic tests in parallel, it is possible to determine a stress value, under which an intensive process of crack formation begins in concrete of specified composition.

**Selection of non-destructive method** The advantage of ultrasonic (high-frequency) waves is the possibility to identify small defects as the wave length is comparable to sizes of defects. However, as they are short, the ultrasonic waves are subject, in a greater degree, to loss of energy due to internal friction (attenuation) and reflection from non-homogeneous medium (dispersion).

The advantage of low-frequency (long) waves is smaller attenuation, due to which they propagate to longer distances. That is why the seismic-acoustic method is used to control a length of piles and to identify major defects.

## Testing of the piles under static load

Check tests of piles under static load are carried out to verify the possibility of transferring to the piles the loads specified in the design. In other words, the objective of carrying out the check tests is to verify the fulfillment of the following condition

$$N \leq \frac{F_d}{\gamma_k}, \tag{4.13}$$

where $N$ = load on pile according to design; $F_d$ = bearing capacity of pile soil base determined based on test data; $\gamma_k$ = safety factor, which is taken as equal 1.2 in determining $F_d$ based on results of field tests under static load.

Piles subject to check tests shall be indicated in the foundation design on the pile field plan. Thus, a piling contractor meets all technological requirements when constructing the above piles.

It is appropriate to test those piles, the workmanship of which causes concern after they have been tested by the non-destructive methods. In some instances, it might be difficult due to insufficient quantity of anchor piles and impossibility to install thrust structures. The problem can be solved by using a loading platform with weights that serve as a jacking pad.

**Testing of piles by standard method.** The check tests of piles under static pressing, pulling or horizontal loads shall be carried out in accordance with the requirements of, GOST 5686, SNiP 2.02.03 and SP 50-102.

According to TSN 50-302, a number of tested piles of every type shall be at least 2 each. A type of pile is determined based on length, diameter and construction technology.

*Equipment and instrumentation.* For the purpose of the pile testing, the test facilities are used that include the following:

- arrangement to apply load to piles (jacks or calibrated weight);
- thrust structure to bear reaction forces (framing or trussing system with anchor piles and/or loading platform);
- arrangement to measure displacements of pile during tests (benchmark system with instrumentation).

For the tests of piles under static pressing load, the most widely used are the installations with one jack or a group of jacks rested on a supporting frame (made of beams or trusses) linked to anchor piles (Figure 4.12). When the reaction pulling forces cannot be completely taken up by the anchor piles, the jack is rested on the platform where the calibrated weight is placed. The pile jacking rig made by Glavleningradstroy Trust No. 101 can be used as loading platform.

The installation shall be designed to bear a load of 20 % higher than the load provided for by the test program.

To eliminate the effect of the anchor piles on a pile to be tested when carrying out the check tests, one should follow the considerations given below. A tip of the anchor piles shall be positioned no lower than a tip of the tested pile. A distance from the longitudinal axis of the tested pile to the anchor pile, as well as to the carriers of the benchmark system shall be at least three times of the largest dimensions of the pile cross section (up to 800 mm in diameter) but no shorter than 1.5 m. For the piles larger than 800 mm in diameter, as well as for the screw piles, the clear distance between the tested and anchor piles can be taken as, at least, two times of the largest dimensions of the pile cross section.

The pile displacements are measured by means of two or more instruments (deflectometers or indicators) that provide a measurement error of no more than 0.1 mm. The instruments for the displacement measurement shall be symmetrically installed at the equal distances (no longer than 2 m) from the tested pile.

All instruments used to measure the pile displacements and loads shall be calibrated and checked against the rated values at regular intervals. Prior to shipping them to the test location, the instruments shall be subject to exceptional verification.

Before testing the piles under static pressing load, the anchor piles are to be prepared in accordance with a planned way of bearing the reaction pulling forces: through the lateral friction (Figure 4.12) or the exposed reinforcing bars of the piles (Figure 4.13).

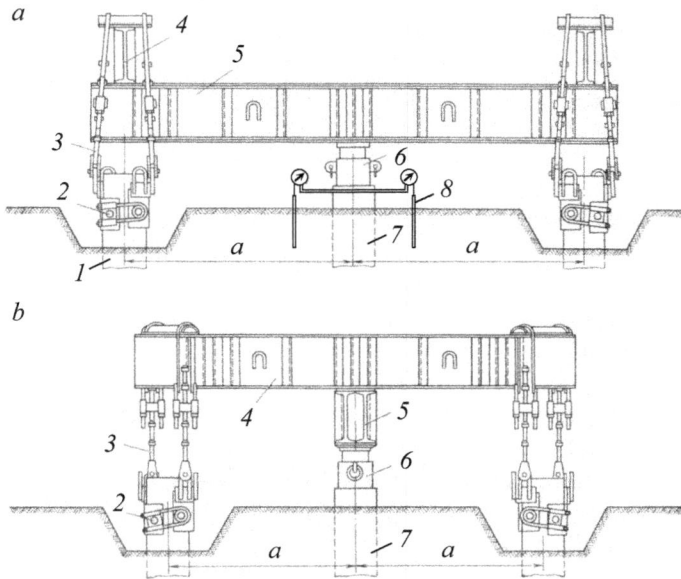

Figure 4.12. Schematic diagram of installation to test factory-made prismatic
piles under static pressing load:
*a* – front view; *b* – side view (benchmark system is not shown for clarity);
*1* – anchor pile; *2* – clamp; *3* – tie rod; *4* – embedded beam; *5* – jacking beam;
*6* – hydraulic jack; *7* – tested pile; *8* – benchmark system with deflectometers

Figure 4.13. General view of installation to test bored pile
under static pressing load:
*1* – tie rod of Titan anchor pile; *2* – tested pile; *3* – beam of thrust structure

111

*Test procedure.* The testing of driven or jacked piles begins after the piles had a "rest", during which the soil structural bonds restore that were disturbed at the time of their driving.

The "rest" duration depends on the content, properties and conditions of the soil, through which the pile cuts through, and the soils under the bottom end of the pile. For sand soils (except for water-saturated fine and silty soils), the rest time is at least 3 days; and for clay soils, the rest time is at least 6 days.

When cutting through sand soils and when there are rudaceous compact sand or clay soils of hard consistency under the pile tip, the "rest" duration can be reduced to 1 day.

A longer "rest" time is specified when cutting through water-saturated fine and silty sands (at least 10 days), clay soils of soft and very soft consistency (at least 20 days).

The testing of bored and cast-in-place displacement piles is started after the pile concrete has reached 80 % of the design strength. For the piles driven by the other methods, the beginning of the tests shall be established no earlier than 1 day after they has been driven.

The tested pile is loaded uniformly, by increments. An increment of load is taken as equal to no more than 1/10 of the maximum pile load $N_{max}$ specified in the test program. When the bottom ends of the piles penetrate rudaceous soils, gravel and compact sands and clay soils of hard consistency, the first three increments of load are to be taken as equal to 1/5 of the maximum load.

At every increment of the pile load, readings shall be taken by all instruments to measure deformation in the following sequence: before loading the pile (zero reading), immediately after the application of the load (first reading), then, in a sequential order, four readings at 30-minute intervals and further every hour till the conventional stabilization of deformation (displacement attenuation).

As a criterion of the conventional deformation stabilization, a rate of pile settlement development is taken at a given increment of load that does not exceed 0.1 mm for the last:

60 min of monitoring if sand soils or clay soils of hard to tough consistency occur under the bottom end of the pile;

2 hours of monitoring if clay soils of soft to flow consistency occur under the bottom end of the pile;

When testing a pile, a load shall be brought to a value, at which the total pile settlement would be at least 40 mm.

When the bottom ends of the piles penetrate rudaceous compact sand or clay soils of hard consistency, a load shall be brought to a value that is as much as 1.5 times the bearing capacity of the pile calculated in accordance with SNiP 2.02.03 or SP 50-102.

112

During the check test of the pile in any case, the maximum load $N_{max}$ shall not be higher than the material-specific bearing capacity of the pile $F_m$.

Upon reaching the maximum load, the pile is unloaded by decrements. Decrements of load are taken to be as much as two times increments of load. Every decrement of load shall be held at least 15 min. To measure deformation, the instrument readings are to be taken immediately after every decrement of load and every 15 min of monitoring.

After the full unloading (down to zero), the elastic displacement of the pile is monitored for 30 min in case of sand soils occurring under the bottom end of the pile and 60 min in case of clay soils. The instrument readings are to be taken every 15 min.

A test logbook shall be maintained during the tests.

*Test results* are presented in the form of the graphs of pile deformation (settlement, withdrawal, horizontal displacement) as a function of load and the graphs of deformation change with time by increments of load (Figure 4.14).

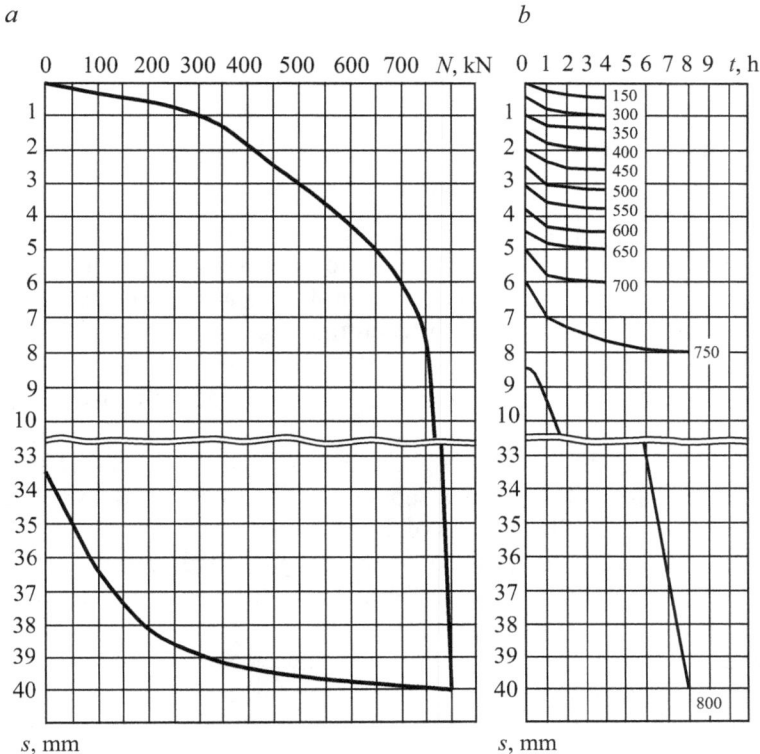

Figure 4.14. Graphic presentation of test results of piles under static pressing load:
*a* – graph of pile settlement *s* as a function of load *N*; *b* – graphs of pile settlement development *s* with time *t* for every increment of load

The appearance of the graph of settlement $s$ as a function of load $N$ is determined by soils that are cut through by the piles and, in particular, by those occurring under the bottom ends (Figure 4.15).

If, in the course of testing the piles under the static load $N_{max} = F_m$, a graph of Type $1$ is obtained (Figure. 4.15, $a$), this implies that the bearing capacity of the soil base of the piles (soil-specific bearing capacity of pile) exceeds the material-specific bearing capacity of pile. That is possible when the pile tip penetrates low-compressible soils (rudaceous soils; gravel and compact sands; clay soils of hard consistency).

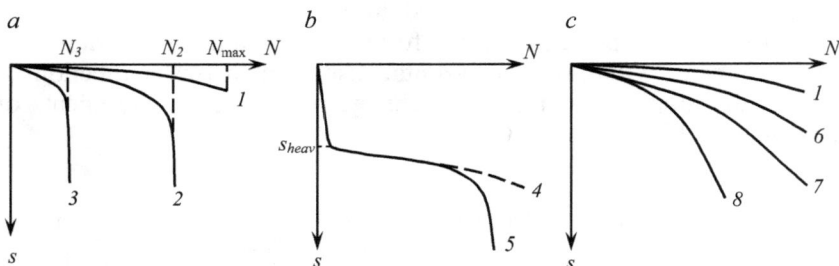

Figure 4.15. Typical appearance of graphs of pile settlement $s$
as a function of load $N$

A graph of Type $2$ (Figure 4.15, $a$) can be obtained as a result of the static testing of the pile under load $N_{max} = F_m$. According to this graph, there is a continuous increase in deformation (pile failure) at a certain load $N_2$, which gives evidence that the soil base of the pile has no bearing capacity left. That is, in this case, the soil-specific bearing capacity of the pile is lower than the material-specific bearing capacity of the pile.

When the pile tip is not brought to the superface of a bearing stratum, the graph of pile settlement as a function of load will be of Type $3$ (Figure 4.15, $a$) instead of the graphs of Type $1$ or $2$.

The consecutive penetration of driven, jacked or cast-in-place displacement piles at close distance from each other can result in the heaving of some already buried piles. When testing a pile that was raised by heaving, the graph $s = f(N)$ has an appearance of Type $4$ or $5$ (Figure 4.15, $b$) depending on the compressibility of the bearing stratum. The application of the first increment of load is accompanied by a caving settlement $s_{heav}$ of the pile by a value of its heaving. After that, there is a stage of soil compacting under the pile foot.

*Interpretation of test results.* The actual bearing capacity of a pile is calculated by the following formula:

$$F_d = \gamma_c \frac{F_{u,n}}{\gamma_g}, \tag{4.14}$$

where $\gamma_c$ = factor determining the pile behavior; in case of pressing or horizontal loads, $\gamma_c = 1$; $F_{u,n}$ = rated value of pile ultimate strength, kN; $\gamma_g$ = soil-specific safety factor.

The standard value of the pile ultimate strength $F_{u,n}$ and the value of soil-specific safety factor $\gamma_g$ are determined based on the results of statistical analysis for particular values of pile ultimate strength $F_u$, following the requirements of GOST 20522 as applied to the procedure given in that document to determine the temporary resistance at the confidence probability $\alpha = 0.95$.

If fewer than six piles of the same type were tested in the same soil conditions, then the standard value of the pile ultimate strength $F_{u,n}$ is to be taken as equal to the minimum ultimate strength $F_{u,\min}$, which was obtained from the test results, and the soil-specific safety factor is to be taken as $\gamma_g = 1$.

The particular value of the soil-specific pile ultimate strength $F_u$ is to be taken as load $N$, at the constant value of which the continuous increase in settlement $s$ (pile failure) is observed.

If the pile was not brought to the failure (graphs of Type $1$, $6$, $7$ and $8$ in Figure 4.15, $c$), determining the pile ultimate strength $F_u$ is arbitrary as it is not clear to which settlement it corresponds. Different criteria are used to estimate $F_u$ by graphs of Type $1$, $6$, $7$ and $8$. For instance, SP 50-102 recommends to take the soil-specific pile ultimate strength $F_u$ as load $N$, under action of which the pile would have a settlement equal to $s$ and which is to be determined by the following formula:

$$s = \zeta \, \overline{s}_u, \tag{4.15}$$

where $\overline{s}_u$ = limit value of average settlement of foundation of structure under construction, and this limit value is to be determined according to SNiP 2.02.01; $\zeta$ = coefficient of transition from limit value of average settlement of foundation of structure $\overline{s}_u$ to pile settlement obtained in static tests at conventional stabilization (attenuation) of settlement.

In accordance with SP 50-102, a value of coefficient $\zeta$ should be taken as equal to 0.2 in cases when the piles are tested at the conventional stabilization equal to 0.1 mm for 1 hour, if sand soils or clay soils of hard to tough consistency occur under their bottom ends, as well as for 2 hours, if clay soils of soft to flow consistency occur under their bottom ends. It is allowed specifying a value of coefficient $\zeta$ by the results of the settlement monitoring in buildings constructed on pile foundations in similar soil conditions.

If the settlement determined by Formula (4.15) proves to be greater than 40 mm, then the standard value of the pile ultimate strength $F_u$, according to SP 50-102, should be taken as a load corresponding to $s = 40$ mm.

If, at the maximum test load $N_{max}$, which will prove to be equal to or higher than 1.5 $F_d$ (where $F_d$ = bearing capacity of pile determined by computational method), the pile settlement $s$ will turn out to be, during the tests, smaller than the value determined by Formula (4.15), then, in this case, SP 50-102 allows taking the standard value of the pile ultimate strength $F_u$ as the maximum load $N_{max}$ obtained during the tests.

**Testing of piles by accelerated method** Being obtained from the field pile tests on a site with soils of homogeneous strike at different application rates of constant-sign monotonic and cyclic loads, the test results allowed establishing that the load value $F_u$, under which a "failure" occurs, is the same for the piles that are in the same conditions, and does not depend on the settlement rate [50]. This implies that accelerated tests can be carried out to determine $F_u$. Notwithstanding that the load – settlement curves are different for different settlement rates, all these curves would have the common asymptote $F = F_u$. This fact was previously noted in the publications by A. A. Luga.

At single monotonic loading, every loading rate has its own corresponding load – settlement curve. In case of constant-sign cyclic loading, an enveloping curve can be plotted that always coincides with a curve of single monotonic loading. The latter curve is rather stable for a particular pile, that is, it does not depend on a number of cycles, but is only determined by the rate of loading and has even more stable vertical asymptote $F = F_u$, which is the same for all such curves and does not depend on a number of completed loading cycles.

A comparison to the curves resulted from the tests with the settlement stabilization at every increment showed that the piles have the same ultimate load $F_u$ as at the constant settlement rate.

**The pile testing by the Osterberg method** is carried out at the pre-design stage, that is, prior to pile designing and large-scale driving. The method allows`separately determining the soil bearing capacity on pile tip and on pile side surface [45, 46]. This method is usually used for testing bored or cast-in-place displacement piles of large dimensions.

The specific feature of the method lies in using for loading an Osterberg cell, which is a powerful jack enclosed in a protective casing. The cell is placed directly on ground at the borehole bottom or on the surface of concrete that fills the borehole in part (Figure 4.16).

Prior to lowering it, supporting slabs are welded to the sell, with the diameter of the supporting slabs is slightly smaller than the borehole diameter. Displacement converters are installed between the slabs. The displacement converters are used to measure an expansion (increase in height) of the Osterberg cell when loading a pile. To install the cell at a certain depth, one of its supporting slabs is welded to the end of the rein-forcing cage. On the reinforcing cage, sensors are installed to record stresses in the pile shaft during the tests.

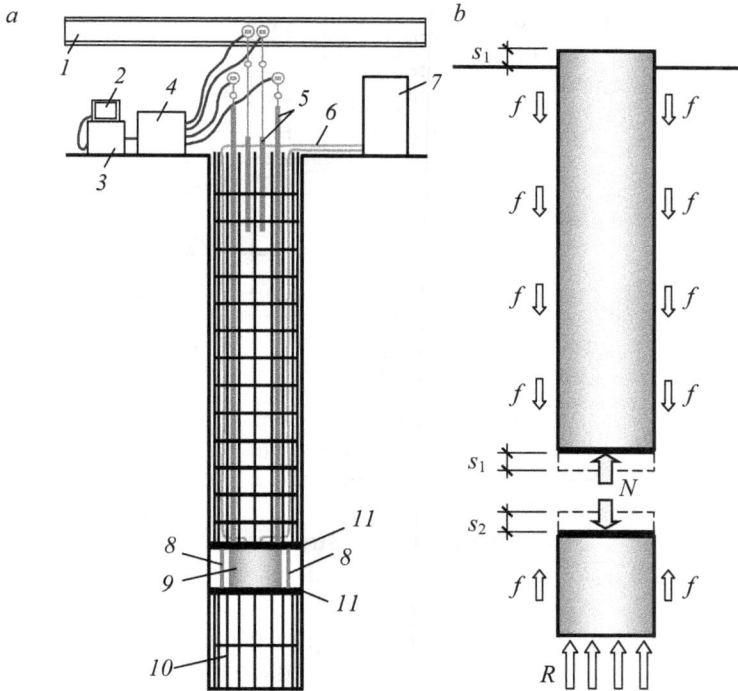

Figure 4.16. Pile testing by Osterberg method:
$a$ – schematic diagram; $b$ – computational model; $1$ – beam of benchmark system; $2$ – computer; $3$ – data recording device; $4$ – switch gear; $5$ – displacement trans-ducer; $6$ – hydraulic system mains; $7$ – hydraulic system control station; $8$ – dis-placement converter; $9$ – Osterberg cell; $10$ – reinforcing cage; $11$ – distribution slab; $f$ and $R$ = soil resistance on pile side surface and under tip; $N$ = load trans-ferred by Osterberg cell on upper and lower part of pile; $s_1$ and $s_2$ = vertical dis-placements of upper and lower part of pile, respectively

The tests are started after the pile concrete has reached 80 % of the design strength and the benchmark system has been installed to measure vertical pile displacements.

The pile is loaded by increments using the Osterberg cell. Every in-crement of load is held till the conventional stabilization of deformation. During the tests, the load is recorded that is transferred to the pile by means of the Osterberg cell, and the measurements of the pile withdrawal

$s_1$ and the cell expansion $s = s_1 + s_2$ are taken. The pile withdrawal is measured using the benchmark system installed over the pile. The benchmark system includes stands and crossbars, to which the displacement transducers are attached.

The results of the static tests are presented in the form of graphs of vertical displacement as a function of load (Figure 4.17) and time. The bearing capacity of the pile soil base is determined by the procedure described above.

When testing long piles, it is required to take into consideration the vertical deformation of their shafts. For this purpose, stress sensors are fastened at certain spacing on the reinforcing cages. The stress values obtained during the pile testing are used in the office processing to calculate the shaft deformation.

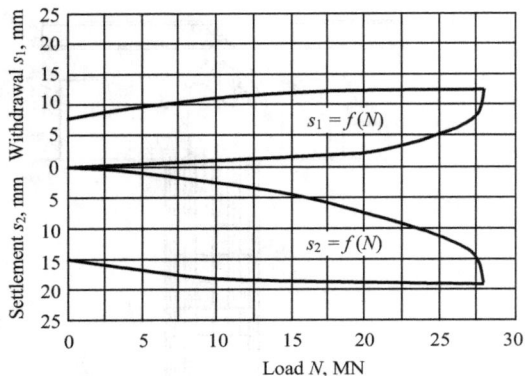

Figure 4.17. Graphs of vertical displacement as a function of load $N$ (graphs were obtained in testing piles by Osterberg method)

The Osterberg method allows testing large-size piles without using anchor piles, which makes it possible to reduce the costs at the stage of geotechnical survey.

To test factory-made piles driven by static or dynamic method, there are used dedicated piles with internal passage, and the Osterberg cells are mounted on the bottom ends of the piles by means of special-purpose fastenings.

**Testing of piles by Van Weel's method.** The method allows dividing the total soil resistance into the resistance under pile tip $R$ and the resistance on pile side surface $f$ [49]. For that, a pile is tested under static load. When testing after application of each of the loads, the piles are completely extracted while taking measurement of pile elastic settlement. The test result is presented in the form of a "load – elastic settlement" graph (Figure 4.18). Many tests carried out by this procedure showed that, after the full use of friction forces on the pile side surface, the graph has a form of a straight line. A straight line drawn from the center of coordinates in parallel to the above straight line characterizes the soil resistance under the pile tip. In Figure 4.18, the crosses indicate the soil resistance under the pile tip determined by the direct measurement using resistance tensiometers [44] As it can be seen from this figure, the straight line drawn from the center of coordinates in parallel to the end section of the "load – elastic settlement" curve shows rather accurately the soil resistance under the pile tip at different elastic settlements of the pile. Thus, this line divides the total pile resistance into the soil resistance under the tip and the resistance on the side surface.

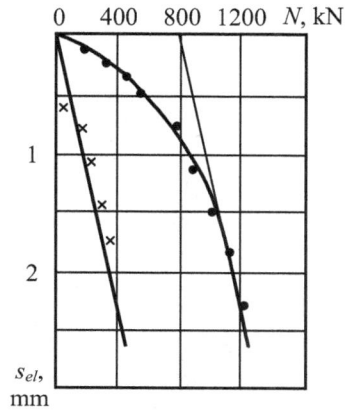

Figure 4.18. "Load – elastic settlement"graph

When plotting a graph similar to the one shown in Figure 4.18, it is required, when determining the elastic pile settlement, to take into consideration its elastic deformation that depends on the distribution of a load applied to the pile into the soil resistance under the tip and on the side surface. Therefore, the graph is to be plotted by the successive approximation method. At a first approximation, no consideration is given for the elastic deformation of the pile and a "load – elastic settlement" graph is plotted. Using the graph, loads are determined as applied to the tip $F_R$ and to the side surface $F_f$, based on which the elastic deformation of the pile is found at every load. The elastic deformation of the pile is approximately determined by the following formula:

$$s_{el} = \frac{l}{A\,E}\left(F_R + \frac{F_f}{2}\right),\qquad(4.16)$$

where $A$ and $l$ = cross sectional area of pile and its length, respectively; $E$ = modulus of elasticity of pile material; $F_R$ and $F_f$ = loads applied to tip and pile side surface, respectively.

A graph is plotted at second approximation, but this time, taking into consideration the elastic deformation of the pile.

# CONCLUSION

The building boom began in the major Russian cities from the middle of 1990s and resulted in the necessity to implement and widely use new types of pile foundations including those made of piles constructed in the ground. Construction companies, in mass order, started to modify drilling rigs as the equipment for the construction of cast-in-situ, drilled and drilled injected piles. Major companies began to purchase or lease new or used rigs of foreign production.

In its turn, that required the development of the technologies of pile construction and driving - in many ways, new for the domestic foundation construction - and their adaptation to the local soil conditions.

Being based on the recommendations of the domestic construction regulations, the calculations by far not always provide a reliable estimate of the bearing capacity and deformation of pile foundations constructed using the new technologies. It can be explained by the fact that the technical standards and rules, which were developed back at the times of the Soviet Union, were primarily oriented to prefabricated piles lowered by driving or to some types of drilled cast-in-situ piles. It is for the above types of piles that numerous scientific researches were carried out, and the results of these researches were used as a basis for the regulatory tables to determine the bearing capacity of pile base, and those regulatory tables exist up to now.

In view of that, a considerable difference can be observed rather frequently between the calculated and actual bearing capacity and settlement of piles. In many instances, a reason for such difference is insufficient study of the physical processes occurring in the soil body during the pile construction by state-of-the-art technologies.

The development of reliable methods to estimate the interaction of such piles with the soil base requires integrated scientific studies.

At the moment, the problems of current interest for geotechnical engineers are the adaptation of the new pile construction technologies to the local soil conditions and the development of the reliable methods to estimate their bearing capacity.

# References

1. I.A. Ganitchev. Construction of man-made bases and foundations / I. A. Ganitchev. – Moscow: Stroyizdat Publishing House, 1981. – 543 pages.
2. L.M. Glozman. Dynamic monitoring in construction of deep foundations in conditions of Saint Petersburg / L.M. Glozman // Third International Geotechnical Conference dedicated to 300th anniversary of Saint Petersburg: in 2 volumes – Saint Petersburg, Moscow: Publishing House of Association of Construction Higher Education Institutions, 2003. – Volume 2. – Pages 67–72.
3. B.I. Dalmatov. Experience of construction in soft soil (by example of construction of Transportation and Commercial Center in Saint Petersburg) / B.I. Dalmatov // Urban reconstruction and geotechnical construction. – Saint Petersburg, 1999. – No. 1. – Pages 4-7.
4. D.E. Dolidze. Testing of structures and facilities / D. E. Dolidze. – Moscow: Higher School Publishing House, 1975. – 252 pages.
5. A.N. Dranovsky. Drilled cast-in-situ piles and foundations of diaphragm-wall type in difficult engineering and geological conditions: Manual / A.N. Dranovsky, O. P. Kalashnikova; edited by. M. T. Kuleyeva. – Kazan: KKhTI Publishing House, 1985. – 80 pages.
6. V.N. Zhelezkov. Screw piles in power industry and other construction industries / V.N. Zhelezkov. – Saint Petersburg: Pragma Publishing House, 2004. – 126 pages.
7. P.A. Konovalov. Bases and foundations of reconstructed buildings / P. A. Konovalov. – Moscow: VNIINTPI, 2000. – 320 pages.
8. D.S. Konyukhov. Construction of urban underground shallow-foundation facilities. Special-purpose works / D. S. Konyukhov. – Moscow: Architecture-S Publishing House, 2005. – 304 pages.
9. Practical guidelines for construction of enclosures made of drilled-cast-in-situ piles. – Moscow: PKTIPromstroy Publishing House, 2001.
10. M.I. Nikitenko. И. Drilled injected anchors and piles in construction and reconstruction of buildings and facilities / M. I. Nikitenko. – Minsk: BNTU Publishing House, 2007. – 580 pages.
11. New ways in anchor technique: ISCHEBECK TITAN anchor piles. Designing and calculation. Störring Satz + Druck GmbH, Ennepetal, 2004 г. – 32 pages.
12. E.M. Perley M. Reinforced concrete pipe piles and shafts-shells for industrial and public construction/ E. M. Perley, N. Ya. Tsukerman. – Leningrad: Stroyizdat Publishing House, 1969. – 200 pages.
13. E.M. Perley M. Pile foundations and underground structures in reconstruction of operating production facilities / E. M. Perley, V. F. Rayuk, V. V. Belenkaya, A. N. Almazov. – Leningrad: Stroyizdat Publishing House, 1989. – 176 pages.
14. Recommendations for application of drilled cast-in-situ piles. – Moscow: NIIOSP Publishing House, 1984.
15. Recommendations for designing and construction of cast-in-situ piles in reamed boreholes. – Moscow: NIIOSP Publishing House, 2000.
16. Piles and pile foundations: reference book / N. S. Metelyuk [et al.]. – Kiev: Budiveknik, 1977. – 256 pages.
17. Piling works / I. I. Kosorukov [et al.]; edited by I. I. Kosorukov. – Moscow: Higher School Publishing House, 1974. – 391 pages.
18. Piling works: builder's reference book / M. I. Kosorukov [et al.]; edited by M. I. Smorodinov. – Moscow: Stroyizdat Publishing House, 1988. – 224 pages.
19. M. I. Smorodinov. Anchor arrangements in construction / M. I. Smorodinov. – Moscow: Stroyizdat Publishing House, 1983. – 182 pages.

20. S.A. Ter-Galustov. Deep drilling supports / S.A. Ter-Galustov. – Moscow: Publishing House of Public Utilities Ministry of RSFSR, 1961. – 128 pages.
21. T.M. Stoll. Construction technology of undestructures for buildings and facilities / T.M. Stoll, V. I. Stoll, V. I. Feklin. – Moscow: Stroyizdat Publishing House, 1990. – 288 pages.
22. A.M. Yagudin. Drilled cast-in-situ piles with onion-shaped enlargements / A.M. Yagudin. – Saratov: Publishing House of Saratov State University, 1983. – 168 pages.
23. VSN 16–84. Instruction on underpinning of foundations of failure-condition and reconstructed buildings using multisection piles.
24. VSN 309–84. Designing and construction of cast-in-situ piles using vibratory technology.
25. VSN 490–87. Designing and construction of pile foundations and sheet piling in conditions of reconstruction of production facilities and urban development.
26. GOST 5686-94. "Soils. Field test methods using piles".
27. GOST 10180-90. "Concrete. Methods for determining of strength by reference samples."
28. GOST 10181-2000. "Concrete mixtures. Test methods".
29. GOST 14098–91. "Welding joints of reinforcement bars and embedded parts of reinforced concrete structural units. Types, designs and dimensions."
30. GOST 17624-87. "Concrete. Ultrasonic method for determining of strength".
31. GOST 18105-86. "Concrete. Rules for strength control."
32. GOST 20522-96. "Soils. Methods for statistic processing of test results".
33. GOST 22685–89. "Molds for making of concrete reference samples. Technical specifications".
34. GOST 22690-88. "Concrete. Determining of strength by non-destructive mechanical methods."
35. GOST 28570-90. "Concrete. Methods for determining of strength by samples taken from structures".
36. RTM 36.44.12.2–90. "Designing and construction of foundations made of piles driven by jacking methods."
37. SNiP 2.02.03–85 "Pile foundations".
38. SNiP 3.03.01-87 "Bearing and enclosing structures"
39. SP 50-102–2003 "Designing and construction of pile foundations".
40. TR 50-180–06 "Technical recommendations for designing and construction of pile foundations made by using pulse-discharge technology for high-risers (PDT piles)".
41. TSN 50-302–2004 "Designing of foundations for buildings and facilities in Saint Petersburg".
42. Amir, J. M. Wave Velocity in Young Concrete / J. M. Amir // Proc. 3rd Int. Conf. on Application of Stress Wave Theory to Piles. – Saint Petersburg, 1988. – P. 911, -912.
43. Belgian screw pile technology: design and recent developments / J. Maertens, N. Huybrechts (eds). – Lisse: Swets & Zeitlinger, 2003. – 372 pages.
44. Mohan, D. Load-bearing capacity of piles / D. Mohan, G. Jain, V. Kumar // Geotechnique, 1963. – Vol. 13, № 1. – P.76–86.
45. Osterberg, J. O. New device for load testing driven piles and drilled shafts separates friction and end bearing / J. O. Amir // Proc. Int. Conf. on Piling and Deep Foundations. – London: Balkema, 1989. – P. 421–431.
46. Osterberg, J. O. The Osterberg load test method for bored and driven piles. The first ten years / J. O. Amir // Proc. 7rd Int. Conf. and Exhibition on Piling and Deep Foundations, Vienna, Austria, June 1998. – Deep Foundation Institute, Englewood Cliffs, New Jersey, 1998. – P. 1.28.1–1.28.11.

47. Screw piles: installation and design in stiff clay / A. E. Holeyman (ed.). – Lisse: Swets & Zeitlinger, 2001. – 338 pages.
48. Tomlinson, M. J. Pile design and construction practice / M. J. Tomlinson. – Abingdon: Taylor & Fransis, 1994. – 411 p.
49. Van Weel, A. F. A Method of separating the bearing capacity of a test pile into skin-friction and point-resistance/ A. F. Van Weel // Proc. 4$^{rd}$ Int. Conf. Soil Mech. and Found. Eng. – London, 1957. – Vol. 2. – P. 76-80.
50. Whitaker T. A. New Approach to Pile Testing / T. A. Whitaker, R. W. Cooke // Proc. 5$^{rd}$ Int. Conf. Soil Mech. and Found. Eng. – Paris, 1961. – P. 171-176.
51. Patent US 4458765, Int. Cl. E21B 7/26. Tool for forming a hole in macroporous compressible soil / V.I. Feklin, A.N. Mironenko, S.V. Shatov, N.S. Shvets, J.A. Kirichek (SU). – Appl. No: 377684; Filed: May 12, 1982; Date of Patent: Jul. 10, 1984.
52. Patent US 4484640, Int. Cl. E21B 7/26. Tool for forming of holes in macroporous compressible soils / V.I. Feklin, V.B. Shvets, B.M. Mazo (SU). – Appl. No: 397438; Filed: Jul. 12, 1982; Date of Patent: Nov. 27, 1984.
53. Patent US 4496011, Int. Cl. E21B 7/26. Tool for forming earth holes having fixed walls and method therefor / B.M. Mazo, V.I. Feklin (SU). – Appl. No: 402073; Filed: Jul. 26, 1982; Date of Patent: Jan. 29, 1985.
54. Patent US 4504173, Int. Cl. E02D 5/56. Apparatus for constructing cast in place tubular piles and method of constructing such piles by same apparatus / V.I. Feklin (SU). – Appl. No: 421090; Filed: Sep. 22, 1982; Date of Patent: Mar. 12, 1985.
55. Patent US 4623025, Int. Cl. E02D 5/36. Soil-displacement drill and method for manufacturing a pile / A.J. Verstraeten (BE). – Appl. No: 659790; Filed: Oct. 11, 1984; Date of Patent: Nov. 18, 1986.
56. Patent US 5722498, Int. Cl. F21B 7/26. Soil displacement auger head for installing piles in the soil / W.F. Van Impe, G.A.A. Cortvrindt (BE). – Appl. No: 637747; Filed: Oct. 28, 1994; Date of Patent: Mar. 3, 1998
57. Patent US 6033152, Int. Cl. E02D 11/00. Pile forming apparatus / K.J. Blum (BE). – Appl. No: 045403; Filed: Mar. 20, 1998; Date of Patent: Mar. 7, 2000.

# Appendix 1

## TECHNICAL FEATURES OF PILE-JACKING RIGS

*Table A.1.1*

### Basic technical features of Russian crawler pile-jacking rigs

| Feature | Identification mark of rig | | | | |
|---|---|---|---|---|---|
| | USV-120 | USV-120M | USV-160 | SUPS-V | SVU-V-6 |
| 1 | 2 | 3 | 4 | 5 | 6 |
| Basic machine | Excavator EO-6122 | Excavator EO-6122 | Excavator EO-6122 | Excavator EO-6122 | Crane RDK-250 |
| Type of jacking mechanism | Hydraulic | Hydraulic | Hydraulic | Hydraulic | Sheave-block |
| Maximum jacking force without weight, kN | 800 | 800 | 800 | 800 | 400 |
| Maximum jacking force with weight, kN | 1200 | 1200 | 1600 | 1600 | 900 |
| Jacking speed, m/min | No higher than 2.0 | No higher than 2.0 | No higher than 2.0 | No higher than 2.0 | 0.5...2.5 |
| Cross sections of driven piles: square cross section, cm | 30×30; 35×35; 40×40 | 30×30; 35×35; 40×40 | 30×30; 35×35; 40×40 | 30×30; 35×35; 40×40 | No higher than 35×35 |
| circular cross section, cm | – | – | – | – | No higher than 35 |
| Driven sheet piling | – | Larsen-IV, Larsen-V | – | Larsen-IV, Larsen-V | Any |
| Minimum distance from jacked pile to wall of existing building, m | 1.0 | 1.0 | 1.0 | 1.0 | 0.4 |
| Unladen weight of rig, t | 105 | 117 | 135 | 130 | 112 |

| 1 | 2 | 3 | 4 | 5 | 6 |
|---|---|---|---|---|---|
| Specific pressure of rig on soil, MPa | 0.159 | 0.180 | 0.100 | 0.185 | 0.07 |
| Maximum traveling speed, km/h | 1.5 | 1.5 | 1.5 | 1.5 | 1.0 |
| Maximum power consumption, kW | 150 | 150 | 180 | 90 | 50 |
| Overall dimensions of rig in operating position, m: | | | | | |
| length | 8.8 | 9.5 | 9.9 | 12.0 | 8.6...9.6 |
| height | 13.36* | 16.8* | 5.0 | 5.0 | 20.0 |
| height to cab roof | 3.8 | 3.8 | 3.8 | 3.8 | 3.35 |
| rig width over tracks | 3.8 | 3.8 | 3.8 | 3.8 | 3.3 |
| width over centerlines of side outriggers | 7.55 | 7.55 | 7.55 | 6.5 | 5.2...7.5 |

* With three-section mast

## Basic technical features of Sunward pile-jacking rigs (China)

| Feature | Rig model | | | | | | | | | | | | |
|---|---|---|---|---|---|---|---|---|---|---|---|---|---|
| | ZYJ 80 | ZYJ 120 | ZYJ 180 | ZYJ 240 | ZYJ 320 | ZYJ 420 | ZYJ 500 | ZYJ 600 | ZYJ 680 | ZYJ 800 | ZYJ 900 | ZYJ 1000 | ZYJ 1200 |
| 1 | 2 | 3 | 4 | 5 | 6 | 7 | 8 | 9 | 10 | 11 | 12 | 13 | 14 |
| Maximum jacking force generated by central jacking mechanism, kN | 800 | 1200 | 1800 | 2400 | 3200 | 4200 | 5000 | 6000 | 6800 | 8000 | 9000 | 10000 | 12000 |
| Type of side jacking mechanism | F | F | F | F | F | A | A | A | A | A | A | A | A |

| 1 | 2 | 3 | 4 | 5 | 6 | 7 | 8 | 9 | 10 | 11 | 12 | 13 | 14 |
|---|---|---|---|---|---|---|---|---|----|----|----|----|----|
| Maximum jacking force generated by side jacking mechanism, kN | 560 | 840 | 1260 | 1680 | 2240 | 1680 | 2000 | 2400 | 2720 | 3200 | 3600 | 4000 | 4800 |
| Dimensions of square piles, cm | 20...30 | 15...35 | 20...40 | 25...50 | 25...50 | 30...55 | 30...55 | 30...55 | 30...55 | 30...55 | 30...50 | 30...50 | 50...70 |
| Maximum diameter of circular piles, cm | 30 | 35 | 40 | 50 | 50 | 60 | 60 | 60 | 60 | 60 | 80 | 80 | 80 |
| Jacking speed, m/min: minimum | 1.36 | 0.90 | 1.00 | 0.76 | 0.93 | 0.71 | 0.73 | 0.74 | 0.85 | 0.84 | 0.74 | 0.67 | 0.56 |
| maximum | 4.33 | 3.35 | 5.50 | 5.50 | 6.70 | 4.10 | 4.10 | 3.93 | 5.00 | 5.00 | 5.00 | 4.20 | 5.00 |
| Jacking stroke, m | 1.5 | 1.5 | 1.6 | 1.6 | 1.6 | 1.8 | 1.8 | 1.8 | 1.8 | 1.8 | 1.8 | 1.8 | 1.8 |
| Longitudinal pace, m | 1.6 | 1.6 | 2.2 | 3.0 | 3.0 | 3.6 | 3.6 | 3.6 | 3.6 | 3.6 | 3.6 | 3.6 | 3.6 |
| Transverse pace, m | 0.4 | 0.4 | 0.5 | 0.6 | 0.6 | 0.6 | 0.6 | 0.6 | 0.7 | 0.7 | 0.7 | 0.7 | 0.7 |
| Swing angle of rig at a time, degree | 11 | 11 | 8 | 8 | 8 | 8 | 8 | 8 | 8 | 8 | 8 | 8 | 8 |
| Minimum offset from wall, m | 0.45 | 0.50 | 0.80 | 0.80 | 1.00 | 0.68 | 0.68 | 0.68 | 0.68 | 0.68 | 1.00 | 1.00 | 1.00 |
| Minimum offset from corner, m | 0.80 | 0.96 | 1.15 | 1.35 | 1.38 | 1.16 | 1.16 | 1.16 | 1.16 | 1.16 | 1.53 | 1.53 | 1.53 |
| Maximum slope of working area, degree | 11 | 11 | 10 | 10 | 10 | 10 | 10 | 10 | 10 | 10 | 10 | 10 | 10 |
| Piston stroke of hydraulic cylinders (vertical travel), m | 0.65 | 0.65 | 0.75 | 0.90 | 1.00 | 1.00 | 1.00 | 1.10 | 1.10 | 1.10 | 1.10 | 1.10 | 1.10 |

| 1 | 2 | 3 | 4 | 5 | 6 | 7 | 8 | 9 | 10 | 11 | 12 | 13 | 14 |
|---|---|---|---|---|---|---|---|---|----|----|----|----|----|
| Crane lifting capacity, $t$ | 5 | 5 | 8 | 8 | 12 | 12 | 12 | 16 | 16 | 16 | 25 | 25 | 30 |
| Power consumption, kW | 30 | 30 | 59 | 59 | 90 | 90 | 104 | 120 | 141 | 165 | 180 | 180 | 180 |
| Maximum pressure of hydraulic system, MPa | 20.0 | 23.1 | 22.0 | 23.1 | 24.7 | 23.6 | 25.0 | 23.9 | 23.5 | 24.4 | 24.2 | 24.1 | 24.4 |
| Overall dimensions, m: | | | | | | | | | | | | | |
| length | 8.0 | 9.0 | 10.0 | 10.0 | 12.0 | 13.2 | 13.2 | 13.5 | 14.0 | 14.0 | 14.5 | 14.5 | 16.0 |
| width | 4.3 | 4.4 | 5.2 | 6.2 | 6.6 | 7.3 | 7.4 | 7.9 | 8.4 | 8.6 | 9.2 | 9.2 | 9.3 |
| height | 3.0 | 3.0 | 3.0 | 3.1 | 3.2 | 3.1 | 3.1 | 3.2 | 3.2 | 3.2 | 3.3 | 3.3 | 3.3 |
| Unladen weight of rig, t | 82 | 122 | 182 | 245 | 325 | 422 | 502 | 602 | 802 | 802 | 902 | 1002 | 1202 |

*Notes*: 1. In the Table, the following abbreviations are taken for the types of side jacking mechanisms: F = Fixed, A = Attached.
2. The rig height is indicated for the transportation position (with the jacking cylinders removed).

## Basic technical features of Sunward pile-jacking rigs (China)

| Feature | Rig model | | | | | | | | | | | | | |
|---------|---|---|---|---|---|---|---|---|---|---|---|---|---|---|
| | ZYJ 80 | ZYJ 120 | ZYJ 180 | ZYJ 240 | ZYJ 320 | ZYJ 420 | ZYJ 500 | ZYJ 600 | ZYJ 680 | ZYJ 800 | ZYJ 900 | ZYJ 1000 | ZYJ 1200 | |
| 1 | 2 | 3 | 4 | 5 | 6 | 7 | 8 | 9 | 10 | 11 | 12 | 13 | 14 | |
| Maximum jacking force generated by central jacking mechanism, kN | 800 | 1200 | 1800 | 2400 | 3200 | 4200 | 5000 | 6000 | 6800 | 8000 | 9000 | 10000 | 12000 | |

| 1 | 2 | 3 | 4 | 5 | 6 | 7 | 8 | 9 | 10 | 11 | 12 | 13 | 14 |
|---|---|---|---|---|---|---|---|---|---|---|---|---|---|
| Type of side jacking mechanism | F | F | F | F | F | A | A | A | A | A | A | A | A |
| Maximum jacking force generated by side jacking mechanism, kN | 560 | 840 | 1260 | 1680 | 2240 | 1680 | 2000 | 2400 | 2720 | 3200 | 3600 | 4000 | 4800 |
| Dimensions of square piles, cm | 20...30 | 15...35 | 20...40 | 25...50 | 25...50 | 30...55 | 30...55 | 30...55 | 30...55 | 30...55 | 30...50 | 30...50 | 50...70 |
| Maximum diameter of circular piles, cm | 30 | 35 | 40 | 50 | 50 | 60 | 60 | 60 | 60 | 60 | 80 | 80 | 80 |
| Jacking speed, m/min: minimum | 1.36 | 0.90 | 1.00 | 0.76 | 0.93 | 0.71 | 0.73 | 0.74 | 0.85 | 0.84 | 0.74 | 0.67 | 0.56 |
| maximum | 4.33 | 3.35 | 5.50 | 5.50 | 6.70 | 4.10 | 4.10 | 3.93 | 5.00 | 5.00 | 5.00 | 4.20 | 5.00 |
| Jacking stroke, m | 1.5 | 1.5 | 1.6 | 1.6 | 1.6 | 1.8 | 1.8 | 1.8 | 1.8 | 1.8 | 1.8 | 1.8 | 1.8 |
| Longitudinal pace, m | 1.6 | 1.6 | 2.2 | 3.0 | 3.0 | 3.6 | 3.6 | 3.6 | 3.6 | 3.6 | 3.6 | 3.6 | 3.6 |
| Transverse pace, m | 0.4 | 0.4 | 0.5 | 0.6 | 0.6 | 0.6 | 0.6 | 0.6 | 0.7 | 0.7 | 0.7 | 0.7 | 0.7 |
| Swing angle of rig at a time, degree | 11 | 11 | 8 | 8 | 8 | 8 | 8 | 8 | 8 | 8 | 8 | 8 | 8 |
| Minimum offset from wall, m | 0.45 | 0.50 | 0.80 | 0.80 | 1.00 | 0.68 | 0.68 | 0.68 | 0.68 | 0.68 | 1.00 | 1.00 | 1.00 |
| Minimum offset from corner, m | 0.80 | 0.96 | 1.15 | 1.35 | 1.38 | 1.16 | 1.16 | 1.16 | 1.16 | 1.16 | 1.53 | 1.53 | 1.53 |

| 1 | 2 | 3 | 4 | 5 | 6 | 7 | 8 | 9 | 10 | 11 | 12 | 13 | 14 |
|---|---|---|---|---|---|---|---|---|----|----|----|----|----|
| Maximum slope of working area, degree | 11 | 11 | 10 | 10 | 10 | 10 | 10 | 10 | 10 | 10 | 10 | 10 | 10 |
| Piston stroke of hydraulic cylinders (vertical travel), m | 0.65 | 0.65 | 0.75 | 0.90 | 1.00 | 1.00 | 1.00 | 1.10 | 1.10 | 1.10 | 1.10 | 1.10 | 1.10 |
| Crane lifting capacity, t | 5 | 5 | 8 | 8 | 12 | 12 | 12 | 16 | 16 | 16 | 25 | 25 | 30 |
| Power consumption, kW | 30 | 30 | 59 | 59 | 90 | 90 | 104 | 120 | 141 | 165 | 180 | 180 | 180 |
| Maximum pressure of hydraulic system, MPa | 20.0 | 23.1 | 22.0 | 23.1 | 24.7 | 23.6 | 25.0 | 23.9 | 23.5 | 24.4 | 24.2 | 24.1 | 24.4 |
| Overall dimensions, m: | | | | | | | | | | | | | |
| length | 8.0 | 9.0 | 10.0 | 10.0 | 12.0 | 13.2 | 13.2 | 13.5 | 14.0 | 14.0 | 14.5 | 14.5 | 16.0 |
| width | 4.3 | 4.4 | 5.2 | 6.2 | 6.6 | 7.3 | 7.4 | 7.9 | 8.4 | 8.6 | 9.2 | 9.2 | 9.3 |
| height | 3.0 | 3.0 | 3.0 | 3.1 | 3.2 | 3.1 | 3.1 | 3.2 | 3.2 | 3.2 | 3.3 | 3.3 | 3.3 |
| Unladen weight of rig, t | 82 | 122 | 182 | 245 | 325 | 422 | 502 | 602 | 802 | 802 | 902 | 1002 | 1202 |

*Notes:* 1. In the Table, the following abbreviations are taken for the types of side jacking mechanisms: F = Fixed, A = Attached.
2. The rig height is indicated for the transportation position (with the jacking cylinders removed).

# TECHNICAL FEATURES OF RUSSIAN PILE-SCREWING MACHINES

*Table A.2.1*

### Basic technical features of MZS-219 machine

| Carrier vehicle | KamAZ 53228 |
|---|---|
| Dimensions of screwed pile, m: | |
| blade diameter | Up to 1.2 |
| shaft diameter | Up to 0.4 |
| length of of screwed pile (pile section) | 5.00, 6.35 |
| Maximum pile screwing angle relative to vertical line, degree | 50 |
| Maximum torsion torque of pile screwing, kN·m | 150 |
| Drilling depth, m | 5.7 |
| Drilling diameter, mm | 200, 400 |
| Maximum torsion torque at drilling tool, kN.m | 15 |
| Type of rotation head drive | Mechanical |
| Machine weight, kg | 8300 |
| Overall dimensions in transportation position, mm: | |
| length | 9650 |
| width | 2500 |
| height | 3700 |
| Overall dimensions of machine in operating position, mm: | |
| length | 11400 |
| width | 4500 |
| height | 9400 |
| Unladen weight of machine, kg | 22500 |

*Note.* Ural and KRAZ trucks can be used as carrier machines.

Figure A.2.1. KRAZ-250-based MZS-219 machine

## Basic technical features of UBM-85 and UBM-150 machines

| Feature | Model | |
|---|---|---|
| | UBM-85 | UBM-150 |
| Carrier vehicle | Ural-4320, KamAZ-53228 | Ural-4320, KamAZ-53228 |
| Maximum boom length, m | 12 | 12 |
| Swing angle of string of pipes, degree | 400 | 400 |
| Working sector, degree | 200 | 200 |
| Maximum length of of screwed pile, m | 8 | 8 |
| Pile blade diameter, mm | 500 | 500 |
| Pile shaft diameter, mm | 219 | 219 |
| Maximum screwing depth, m | 24 | 24 |
| Maximum torsion torque of pile screwing, kN·m | 85 | 150 |
| Drilling depth when using continuous flight auger, m | Up to 24 | Up to 24 |
| Drilling diameter when using continuous flight auger, m | 360; 500; 630; 800; 1200; 2000 | 360; 500; 630; 800; 1200; 2000 |
| Maximum torsion torque at drilling tool, kN·m | 15 | 90 |
| Permissible slope of platform, degree | 3 | 3 |
| Lifting capacity of platform, kg | 3000 | 3000 |
| Dead weight of rig, kg | 8300 | 8300 |
| Unladen weight of machine, kg | 17000 | 17000 |

Figure A.2.2. Ural-4320-based UBM-85 machine

# TECHNICAL FEATURES OF RUSSIAN ROTARY DRILLING MACHINES

*Table A.3.1*

### Basic technical features of BKM-1514 and BM-811 machines

| Feature | Machine | |
|---|---|---|
| | BKM-1514 | BM-811 |
| 1 | 2 | 3 |
| Carrier vehicle | KamAZ-53228 | Ural-4320 |
| Type of drilling tool | Short auger | Continuous flight auger |
| Maximum drilling depth, m | 15 | 15 |
| Drilling diameter, mm | 360; 630; 800 | 150; 200; 250; 360; 400; 450; 500 |
| Maximum torsion torque at drilling tool, kN·m | 29.6 | 14.7 |
| Downfeed force of drilling tool, kN | 100 | 100 |
| Upfeed force of drilling tool, kN | 70 | 70 |
| Swing angle of platform, degree | 180 | 180 |
| Rotation speed of drilling tool, rpm | 15...93 | 36...95 |
| Type of drilling tool feed drive | Hydraulic | Hydraulic |
| Type of drilling tool rotary drive | Hydraulic | Hydraulic |
| Type of lifting mechanism drive | Hydraulic | Hydraulic |
| Maximum longitudinal travel of mast, m | 0.8 | 0.8 |
| Capacity of crane equipment, kg | 3000 | 3000 |
| Maximum lifting height of hoist hook, m | 10 | 10 |
| Gradeability in transportation position, degree | | |
| longitudinal | 15 | 15 |
| transverse | 10 | 10 |
| Distribution of load from total machine weight on base, kN: | | |
| through tires of front-axle wheels | 58.8 | 63.8 |
| through tires of rear-bogie wheels | 176.4 | 152.1 |

| 1 | 2 | 3 |
|---|---|---|
| Overall dimensions in transportation position, mm: | | |
|     length | 12300 | 13600 |
|     width | 2500 | 2500 |
|     height | 3560 | 3700 |
| Overall dimensions of machine in operating position, mm: | | |
|     length | 9800 | 10400 |
|     width | 4500 | 4100 |
|     height | 12800 | 13800 |
| Machine weight (total), t | 24 | 22 |

Figure A.3.1. KamAZ-53228-based drilling-and-crane machine BKM-1514

Figure A.3.2. Ural-4320-based drilling machine BM-811

## Basic technical features of MBSh-518 and MBSh-818 machines

| Feature | Model | |
|---|---|---|
| | MBSh-518 | MBSh-818 |
| Carrier vehicle | Ural-4320 | Ural-4320 |
| Drilling with auger: | | |
|     diameter, mm | 500 | 800 |
|     depth, m | 15 | 20 |
| Drilling with hollow auger: | | |
|     diameter, mm | 460 | 460 |
|     depth, m | 20 | 25 |
| Maximum stroke of rotation head, m | 5.3 | 7.5 |
| Maximum torsion torque at drilling tool, kN·m | 14.7 | 40.9 |
| Downfeed force of drilling tool, kN | 60 | 100 |
| Upfeed force of drilling tool, kN | 70 | 70 |
| Traction force of winch, kg | 3000 | 3000 |
| Maximum lifting height of hoist hook, m | 6.5 | 10.0 |
| Rotation speed of drilling tool, rpm | 25...67 | 20 |
| Type of drilling tool feed drive | Hydraulic | Hydraulic |
| Type of drilling tool rotary drive | Hydraulic | Hydraulic |
| Type of lifting mechanism drive | Hydraulic | Hydraulic |
| Gradeability in transportation position, degree | | |
|     longitudinal | 15 | 15 |
|     transverse | 10 | 12 |
| Total weight, kg | 13300 | 21300 |
| Distribution of load from machine on road, kg: | | |
|     through tires of front-axle wheels | 5300 | 5300 |
|     through tires of rear-bogie wheels | 12000 | 16000 |
| Overall dimensions in transportation position, mm: | | |
|     length | 9100 | 10800 |
|     width | 2500 | 2500 |
|     height | 3950 | 3970 |
| Overall dimensions of machine in operating position, mm: | | |
|     length | 9300 | 10510 |
|     width | 2500 | 4500 |
|     height | 8100 | 10820 |
| Maximum traveling speed, km/h | 50 | 50 |

*a*

*b*

Figure A.3.3. Auger-drilling machines MBSh-518 (*a*) and MBSh-818 (*b*)

*Table A.3.3*

**Basic technical features of Modern Drilling Equipment rotary drilling rigs**

| Feature | Drilling rig Sterh SBG-PM2 | Drilling rig UBG-LG1 Alligator | Drilling rig UBG-SA Berkut | Drilling rig UBG-SG Berkut | Drilling-and-crane rig UBKG-TA | Drilling-and-crane rig UBKG-TE |
|---|---|---|---|---|---|---|
| Series | Portable | Light-duty | Medium-duty | Medium-duty | Heavy-duty | Heavy-duty |
| Undercarriage | – | Tracked configuration motorcar configuration or cross-country configuration | Motorcar configuration | Tracked configuration | Motorcar configuration | Tracked configuration |
| Feed force of drilling tool, t | 2 | 5 | 10 | 10 | 25 | 25 |
| Feed force stroke of drilling tool, m | 1.4 | 1.4 | 4.5 | 4.0 | 7.0 | Up to 22.0 |
| Feed speed of drilling tool, m/s | 0...0.9 | 0.11...0.52 | 0.15..0.7 | 0.15...0.7 | 0.15...0.7 | 0.15...0.7 |
| Rotation speed of drilling tool, rpm | 0...380 | 0...200 | 0...51 | 0...51 | 0...53 | 0...53 |
| Maximum torsion torque, kN·m | 2.5 | 5.4 | 35.0 | 40.0 | 80 | 80...150 |
| Maximum drilling diameter, mm | 250 | 320 | 700 | 650 | 1000 | 1000 |
| Drilling depth in soils of Category I-IV using hollow sealed augers, m: | | | | | | |
| ShG-150 (borehole diameter is 150 mm) | 30 | 35 | – | – | – | – |
| ShG-180 (borehole diameter is 180 mm) | 25 | 30 | – | – | – | – |
| ShG-200 (borehole diameter is 200 mm) | 25 | 30 | – | – | – | – |
| ShG-250 (borehole diameter is 250 mm) | 15 | 25 | 40 | 40 | – | – |
| ShG-320 (borehole diameter is 320 mm) | – | 10 | 35 | 35 | – | – |
| ShG-370 (borehole diameter is 370 mm) | – | – | 30 | 30 | 35 | 35 |
| ShG-425 (borehole diameter is 425 mm) | – | – | 25 | 25 | 30 | 30 |
| ShG-450 (borehole diameter is 450 mm) | – | – | 25 | 25 | 30 | 30 |
| ShG-550 (borehole diameter is 550 mm) | – | – | 15 | 15 | 25 | 25 |
| ShG-650 (borehole diameter is 650 mm) | – | – | 10 | 10 | 20 | 20 |
| Total weight, kg | | 1700 | 5000 | 7500 | 22500 | 39500 |

## Range of hollow sealed augers
## Modern Drilling Equipment

| Make | Drilling diameter, mm | Auger diameters, mm | | Auger pipe diameter, mm | Blade thickness, mm |
|------|------|------|------|------|------|
| | | Outside | Inside | | |
| 1 | 2 | 3 | 4 | 5 | 6 |
| ShR-90 | 90 | 80 | 20 | 45 | 4 |
| ShG-90R | | | | | |
| ShG-150 | 150 | 140 | 60 | 89 | 5 |
| ShG-150M | | 145 | | | 12 |
| ShG-150R | | 140 | 63 | 89 | 5 |
| ShG-150RM | | 145 | | | 12 |
| ShG-180 | 180 | 165 | 80 | 114 | 5 |
| ShG-180M | | 170 | | | 12 |
| ShG-180P | | 165 | 80 | 114 | 5 |
| ShG-180PM | | 170 | | | 12 |
| ShG-200 | 200 | 187 | 90 | 127 | 5 |
| ShG-200M | | 190 | | | 12 |
| ShG-200P | | 187 | 90 | 127 | 5 |
| ShG-200PM | | 190 | | | 12 |
| ShG-200R | | 187 | 107 | 127 | 5 |
| ShG-200RM | | 190 | | | 12 |
| ShG-250 | 250 | 230 | 125 | 168 | 5 |
| ShG-250M | | 240 | | | 12 |
| ShG-250P | | 230 | 125 | 168 | 5 |
| ShG-250PM | | 240 | | | 12 |
| ShG-320 | 320 | 300 | 160 | 219 | 8 |
| ShG-320M | | 310 | | | 18 |
| ShG-320P | | 300 | 160 | 219 | 8 |
| ShG-320PM | | 310 | | | 18 |
| ShG-340 | 340 | 320 | 160 | 219 | 8 |
| ShG-340M | | 330 | | | 18 |
| ShG-340P | | 320 | 160 | 219 | 8 |
| ShG-340PM | | 330 | | | 18 |
| ShG-370P | 370 | 350 | 241 | 273 | 8 |
| ShG-370PM | | 360 | | | 18 |
| ShG-425P | 425 | 405 | 293 | 325 | 8 |
| ShG-425PM | | 415 | | | 18 |

| 1 | 2 | 3 | 4 | 5 | 6 |
|---|---|---|---|---|---|
| ShG-450P | 450 | 425 | 293 | 325 | 8 |
| ShG-450PM | | 435 | | | 18 |
| ShG-550P | 550 | 530 | 385 | 426 | 10 |
| ShG-550PM | | 540 | | | 22 |

*Note.* 1. The following notation is taken the table: ShG = sealed auger; R = threaded connection; P = flight type; M = blade with hard-alloyed protection. 2. There are produced augers 1 to 8 m long.

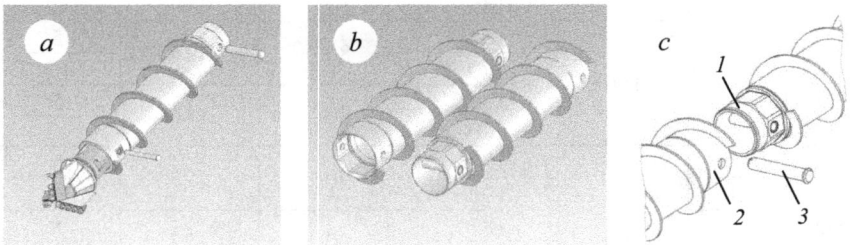

Figure P.3.4. Design of Modern Drilling Equipment hollow sealed augers:
*a* – pilot auger with removable drilling bit; *b* – line augers; *c* – auger joint design;
*1* – union; *2* – coupling; *3* – pin

138

# TECHNICAL FEATURES
# OF BAUER ROTARY DRILLING RIGS

*Table A.4.1*

**Technical features of Bauer rotary drilling rigs**

| Feature | Model | | | | |
|---|---|---|---|---|---|
| | BG 15H BT 40 | BG 18H BT 50 | BG 20H BT 60 | BG 24 BS 70C | BG 24H BS 70C |
| Method of drilling tool feed | Winch | Winch | Winch | Winch | Winch |
| Total height, m | 18.0 | 19.1 | 20.9 | 22.8 | 21.9 |
| Unladen weight, t | 47.5 | 53.0 | 60.0 | 76.5 | 77.5 |
| Torsion torque, kN·m | 151 | 176 | 200 | 233 | 233 |
| Engine capacity, kW | 153 | 187 | 210 | 260 | 260 |
| Traction force of main winch, kN: | | | | | |
| effective | 110 | 140 | 170 | 200 | 200 |
| rated | 140 | 178 | 218 | 256 | 256 |
| Traction force of auxiliary winch, kN: | | | | | |
| effective | 50 | 55 | 55 | 80 | 80 |
| rated | 63 | 68 | 70 | 100 | 100 |
| Winch force at drilling tool feed, kN: | | | | | |
| downward | 100 | 200 | 205 | 200 | 200 |
| upward | 200 | 200 | 210 | 330 | 330 |
| Winch stroke, m: | | | | | |
| Kelly bar | 6.5 | 8.5 | 8.6 | – | 9.0 |
| continuous flight auger | 12.6 | 14.0 | 15.0 | 15.7 | 15.4 |
| Maximum drilling diameter, mm: | | | | | |
| without casing pipe | 1500 | 1500 | 1500 | 2000 | 1700 |
| with casing pipe | 1200 | 1200 | 1200 | 1700 | 1400 |
| Maximum drilling depth, m | 40.7 | 45.5 | 53.3 | 57.6 | 53.8 |
| Angle of lean, degree: | | | | | |
| backward | 15 | 15 | 15 | 15 | 15 |
| forward | 5 | 5 | 5 | 5 | 5 |
| across | 5 | 5 | 5 | 8 | 8 |
| Undercarriage | UW 45 | UW 50 | UW 60 | UW 70 | UW 70 |
| Length of crawler track undercarriage, mm | 4710 | 4750 | 5100 | 5285 | 5285 |
| Width of extendable crawler track undercarriage, mm: | | | | | |
| minimum | 3000 | 3000 | 3000 | 3000 | 3000 |
| maximum | 4000 | 4200 | 4300 | 4400 | 4400 |
| Track width of undercarriage, mm | 600 | 700 | 700 | 700 | 700 |
| Traction force of undercarriage, kN: | | | | | |
| effective | 360 | 350 | 424 | 473 | 473 |
| rated | 420 | 437 | 530 | | |
| Traveling speed, km/h | 1.8 | 1.6 | 1.5 | 1.6 | 1.6 |

| Feature | Model | | | | |
|---|---|---|---|---|---|
| | BG 25C BH 60 | BG 25C BH 60 | BG 28 BS 80B | BG 28 BS 80B | BG 28H BS 80B |
| Method of drilling tool feed | Winch | Hydraulic cylinder | Winch | Hydraulic cylinder | Winch |
| Total height, m | 22.8 | 22.8 | 26.5 | 26.5 | 25.4 |
| Unladen weight, t | 70.0 | 69.0 | 96.0 | 95.0 | 96.0 |
| Torsion torque, kN·m | 245 | 245 | 275 | 275 | 275 |
| Engine capacity, kW | 194 | 194 | 313 | 313 | 313 |
| Traction force of main winch, kN: | | | | | |
| effective | 200 | 200 | 250 | 250 | 250 |
| rated | 250 | 250 | 317 | 317 | 317 |
| Traction force of auxiliary winch, kN: | | | | | |
| effective | 80 | 80 | 80 | 80 | 80 |
| rated | 100 | 100 | 100 | 100 | 100 |
| Winch force at drilling tool feed, kN: | | | | | |
| downward | 260 | 200 | 330 | 250 | 330 |
| upward | 260 | 260 | 330 | 330 | 330 |
| Winch stroke, m: | | | | | |
| Kelly bar | 8.1 | 6.7 | 10.4 | 6.5 | 10.0 |
| continuous flight auger | 15.8 | 15.8 | 19.3 | 17.8 | 18.4 |
| Maximum drilling diameter, mm: | | | | | |
| without casing pipe | 1700 | 1900 | 1900 | 2100 | 1900 |
| with casing pipe | 1400 | 1600 | 1600 | 1800 | 1600 |
| Maximum drilling depth, m | 57.2 | 57.4 | 71.0 | 70.7 | 71.0 |
| Angle of lean, degree: | | | | | |
| backward | 15 | 15 | 15 | 15 | 15 |
| forward | 5 | 5 | 5 | 5 | 5 |
| across | 5 | 5 | 5* | 5* | 8 |
| Undercarriage | UW 65 | UW 65 | UW 95 | UW 95 | UW 95 |
| Length of crawler track undercarriage, mm | 5358 | 5358 | 5680 | 5680 | 5680 |
| Width of extendable crawler track undercarriage, mm: | | | | | |
| minimum | 3200 | 3200 | 3300 | 3300 | 3300 |
| maximum | 4400 | 4400 | 4500 | 4500 | 4500 |
| Track width of undercarriage, mm | 700 | 700 | 800 | 800 | 800 |
| Traction force of undercarriage, kN: | | | | | |
| effective | 422 | 422 | 730 | 730 | 730 |
| rated | 520 | 520 | 860 | 860 | 860 |
| Traveling speed, km/h | 1.5 | 1.5 | 1.1 | 1.1 | 1.1 |

| Feature | Model | | | | |
|---|---|---|---|---|---|
| | BG 36 BS 80B | BG 36 BS 80B | BG 40 BS 100B | BG 40 BS 100B | BG 48 BS 180 |
| Method of drilling tool feed | Winch | Hydraulic cylinder | Winch | Hydraulic cylinder | Hydraulic cylinder |
| Total height, m | 24.2 | 24.2 | 27.1 | 27.1 | 36.0 |
| Unladen weight, t | 120.0 | 115.0 | 140.0 | 139.0 | 250.0 |
| Torsion torque, kN·m | 367 | 367 | 390 | 390 | 482 |
| Engine capacity, kW | 354 | 354 | 433 | 433 | 570 |
| Traction force of main winch, kN: | | | | | |
|    effective | 250 | 250 | 300 | 300 | 600 |
|    rated | 317 | 317 | 384 | 384 | 750 |
| Traction force of auxiliary winch, kN: | | | | | |
|    effective | 100 | 100 | 130 | 130 | 200 |
|    rated | 125 | 125 | 162 | 162 | 256 |
| Winch force at drilling tool feed, kN: | | | | | |
|    downward | 250 | 250 | 460 | 270 | 300 |
|    upward | 400 | 400 | 460 | 400 | 600 |
| Winch stroke, m: | | | | | |
|    Kelly bar | 9.3 | 6.5 | 10.1 | 7.7 | 8.8 |
|    continuous flight auger | 16.7 | 16.4 | 19.7 | 19.4 | – |
| Maximum drilling diameter, mm: | | | | | |
|    without casing pipe | 2300 | 2500 | 2800 | 3000 | 3000 |
|    with casing pipe | 2000 | 2200 | 2500 | 2700 | 2700 |
| Maximum drilling depth, m | 68.6 | 60.1 | 80.5 | 72.1 | 102.2 |
| Angle of lean, degree: | | | | | |
|    backward | 15 | 15 | 15 | 15 | 15 |
|    forward | 5 | 5 | 5 | 5 | 5 |
|    across | 5 | 5 | 5 | 5 | 5 |
| Undercarriage | UW 110 | UW 110 | UW 130 | UW 130 | UW 180 |
| Length of crawler track undercarriage, mm | 5675 | 5675 | 6000 | 6000 | 8800 |
| Width of extendable crawler track undercarriage, mm: | | | | | |
|    minimum | 3400 | 3400 | 3700 | 3700 | – |
|    maximum | 4600 | 4600 | 5000 | 5000 | 7200 |
| Track width of undercarriage, mm | 800 | 800 | 1000 | 1000 | 1200 |
| Traction force of undercarriage, kN: | | | | | |
|    effective | 693 | 693 | 790 | 790 | – |
|    rated | – | – | 930 | 930 | – |
| Traveling speed, km/h | 1.3 | 1.3 | 1.1 | 1.1 | 1.0 |

*Note.* The maximum drilling depth depends on the make of the telescopic Kelly bar (see Table A.4.2).

# Parameters of drilling by Kelly method

| Rig model | Maximum drilling diameter, mm | | $Z_1$, mm | $Z_2$, mm | Make of telescopic bar | $A$, m | $B$, m | Bar weight, t | $H_W$, m | Maximum drilling depth $H$, m |
|---|---|---|---|---|---|---|---|---|---|---|
| | without casing pipe | with casing pipe | | | | | | | | |
| 1 | 2 | 3 | 4 | 5 | 6 | 7 | 8 | 9 | 10 | 11 |
| BG 15H BT 40 Winch | 1500 | 1200 | 3310 | 16000 | BK15/343/3/18 | 8.30 | 20.16 | 3.15 | 5.80 | 18.50 |
| | | | | | BK15/343/3/24 | 10.30 | 26.16 | 3.75 | 3.80 | 24.50 |
| | | | | | BK15/343/3/30 | 12.30 | 32.16 | 4.35 | 1.80 | 30.50 |
| | | | | | BK15/343/4/24 | 8.16 | 26.36 | 3.80 | 5.90 | 24.70 |
| | | | | | BK15/343/4/32 | 10.16 | 34.36 | 4.50 | 3.90 | 32.70 |
| | | | | | BK15/343/4/40 | 12.16 | 42.36 | 5.30 | 1.90 | 40.70 |
| BG 18H BT 50 Winch | 1500 | 1200 | 3240 | 17250 | BK20/368/3/18 | 8.40 | 20.65 | 3.00 | 7.03 | 19.11 |
| | | | | | BK20/368/3/24 | 10.40 | 26.65 | 3.60 | 5.03 | 25.11 |
| | | | | | BK20/368/3/30 | 12.40 | 32.65 | 4.20 | 3.03 | 31.11 |
| | | | | | BK20/368/4/36 | 11.40 | 38.99 | 5.60 | 4.03 | 37.45 |
| | | | | | BK20/368/4/40 | 12.40 | 42.99 | 6.10 | 3.03 | 41.45 |
| | | | | | BK20/368/4/44 | 13.40 | 46.99 | 6.60 | 2.03 | 45.45 |
| BG 20H BT 60 Winch | 1500 | 1200 | 3375 | 18946 | BK20/368/3/24 | 10.40 | 26.65 | 3.80 | 6.65 | 25.00 |
| | | | | | BK20/368/3/30 | 12.40 | 32.65 | 4.60 | 4.65 | 31.00 |
| | | | | | BK20/368/3/36 | 14.40 | 38.65 | 5.30 | 2.65 | 37.00 |
| | | | | | BK20/368/4/32 | 10.40 | 35.00 | 5.20 | 6.65 | 33.30 |
| | | | | | BK20/368/4/44 | 13.40 | 47.00 | 6.70 | 3.65 | 45.30 |
| | | | | | BK20/368/4/52 | 15.40 | 55.00 | 7.70 | 1.65 | 53.30 |

| 1 | 2 | 3 | 4 | 5 | 6 | 7 | 8 | 9 | 10 | 11 |
|---|---|---|---|---|---|---|---|---|---|---|
| BG 24H BS 70C Winch | 1700 | 1400 | 3586 | 19892 | BK25/394/3/24 | 10.71 | 27.20 | 4.70 | 7.10 | 25.50 |
| | | | | | BK25/394/3/27 | 11.71 | 30.20 | 5.12 | 6.10 | 28.50 |
| | | | | | BK25/394/3/30 | 12.71 | 33.20 | 5.53 | 5.10 | 31.50 |
| | | | | | BK25/394/3/36 | 14.71 | 39.20 | 6.35 | 3.10 | 37.50 |
| | | | | | BK25/394/4/32 | 10.71 | 35.47 | 6.85 | 7.10 | 33.80 |
| | | | | | BK25/394/4/36 | 11.71 | 39.47 | 7.15 | 6.10 | 37.80 |
| | | | | | BK25/394/4/40 | 12.71 | 43.47 | 7.73 | 5.10 | 41.80 |
| | | | | | BK25/394/4/48 | 14.71 | 51.47 | 8.85 | 3.10 | 49.80 |
| | | | | | BK25/394/4/52 | 15.71 | 55.47 | 9.45 | 2.10 | 53.80 |
| BG 25HC BH 60 Winch | 1700 | 1400 | 3980 | 20300 | BK 25/394/3/24 | 10.71 | 27.20 | 4.70 | 8.15 | 24.90 |
| | | | | | BK 25/394/3/27 | 11.71 | 30.20 | 5.12 | 7.15 | 27.90 |
| | | | | | BK 25/394/3/30 | 12.71 | 33.20 | 5.53 | 6.15 | 30.90 |
| | | | | | BK 25/394/3/36 | 14.71 | 39.20 | 6.35 | 4.15 | 36.90 |
| | | | | | BK 25/394/4/40 | 12.71 | 43.47 | 7.73 | 6.15 | 41.20 |
| | | | | | BK 25/394/4/48 | 14.71 | 51.47 | 8.85 | 4.15 | 49.20 |
| | | | | | BK 25/394/4/56 | 16.71 | 59.47 | 10.05 | 2.15 | 57.20 |
| BG 25HC BH 60 Cylinder | 1900 | 1600 | 3820 | 18850 | BK 25/394/3/24 | 10.71 | 27.20 | 4.70 | 6.80 | 25.10 |
| | | | | | BK 25/394/3/27 | 11.71 | 30.20 | 5.12 | 6.80 | 28.10 |
| | | | | | BK 25/394/3/30 | 12.71 | 33.20 | 5.53 | 6.15 | 31.10 |
| | | | | | BK 25/394/3/36 | 14.71 | 39.20 | 6.35 | 4.15 | 37.10 |
| | | | | | BK 25/394/4/40 | 12.71 | 43.47 | 7.73 | 6.15 | 41.35 |
| | | | | | BK 25/394/4/48 | 14.71 | 51.47 | 8.85 | 4.15 | 49.35 |
| | | | | | BK 25/394/4/56 | 16.71 | 59.47 | 10.05 | 2.15 | 57.35 |

| 1 | 2 | 3 | 4 | 5 | 6 | 7 | 8 | 9 | 10 | 11 |
|---|---|---|---|---|---|---|---|---|---|---|
| BG 28 BS 80B Winch | 1900 | 1600 | 3780 | 24340 | BK28/419/3/24 | 10.71 | 26.95 | 5.00 | 9.70 | 24.90 |
| | | | | | BK28/419/3/30 | 12.71 | 32.95 | 5.85 | 9.70 | 30.90 |
| | | | | | BK28/419/3/36 | 14.71 | 38.95 | 6.65 | 7.70 | 36.90 |
| | | | | | BK28/419/4/40 | 12.71 | 43.08 | 8.10 | 9.70 | 41.00 |
| | | | | | BK28/419/4/64 | 18.71 | 67.08 | 11.65 | 3.70 | 65.00 |
| | | | | | BK28/419/4/70 | 20.21 | 73.08 | 12.75 | 2.20 | 71.00 |
| BG 28 BS 80B Cylinder | 2100 | 1800 | 4260 | 21420 | BK28/419/3/24 | 10.71 | 26.95 | 5.00 | 6.80 | 24.55 |
| | | | | | BK28/419/3/30 | 12.71 | 32.95 | 5.85 | 6.80 | 30.55 |
| | | | | | BK28/419/3/36 | 14.71 | 38.95 | 6.65 | 6.80 | 36.55 |
| | | | | | BK28/419/4/40 | 12.71 | 43.08 | 8.10 | 6.80 | 40.68 |
| | | | | | BK28/419/4/64 | 18.71 | 67.08 | 11.65 | 3.70 | 64.68 |
| | | | | | BK28/419/4/70 | 20.21 | 73.08 | 12.75 | 2.20 | 70.68 |
| BG 28H BS 80B Winch | 1900 | 1600 | 3670 | 23300 | BK28/419/3/27 | 11.71 | 29.95 | 5.40 | 9.4 | 28.00 |
| | | | | | BK28/419/3/30 | 12.71 | 32.95 | 5.80 | 8.7 | 31.00 |
| | | | | | BK28/419/3/36 | 14.71 | 38.95 | 6.65 | 6.7 | 37.00 |
| | | | | | BK28/419/4/40 | 12.71 | 43.08 | 8.10 | 8.7 | 41.00 |
| | | | | | BK28/419/4/64 | 18.71 | 67.08 | 11.65 | 2.7 | 65.00 |
| | | | | | BK28/419/4/70 | 20.21 | 73.08 | 12.75 | 1.2 | 71.00 |
| BG36 BS 80B Winch | 2300 | 2000 | 3839 | 22063 | BK36/470/3/27 | 12.25 | 30.16 | 7.30 | 7.90 | 28.00 |
| | | | | | BK36/470/3/30 | 13.25 | 33.16 | 7.75 | 6.90 | 31.00 |
| | | | | | BK36/470/3/36 | 15.25 | 39.16 | 8.65 | 4.80 | 37.00 |
| | | | | | BK36/470/3/45 | 18.25 | 48.16 | 10.00 | 1.80 | 46.00 |
| | | | | | BK36/470/4/36 | 12.25 | 38.79 | 9.50 | 7.90 | 36.60 |
| | | | | | BK36/470/4/44 | 14.25 | 46.79 | 11.10 | 5.80 | 44.60 |
| | | | | | BK36/470/4/52 | 16.25 | 54.79 | 12.50 | 3.80 | 52.60 |
| | | | | | BK36/470/4/60 | 18.25 | 62.79 | 13.90 | 1.80 | 60.60 |
| | | | | | BK36/470/4/68 | 20.25 | 70.79 | 15.30 | 1.80 | 68.60 |

| 1 | 2 | 3 | 4 | 5 | 6 | 7 | 8 | 9 | 10 | 11 |
|---|---|---|---|---|---|---|---|---|---|---|
| BG36 BS 80B Cylinder | 2500 | 2200 | 4394 | 21950 | BK36/470/3/27 | 12.25 | 30.16 | 7.30 | 6.80 | 27.50 |
| | | | | | BK36/470/3/30 | 13.25 | 33.16 | 7.75 | 6.80 | 30.50 |
| | | | | | BK36/470/3/36 | 15.25 | 39.16 | 8.65 | 4.80 | 36.50 |
| | | | | | BK36/470/3/45 | 18.25 | 48.16 | 10.00 | 1.80 | 45.50 |
| | | | | | BK36/470/4/36 | 12.25 | 38.79 | 9.50 | 6.80 | 36.10 |
| | | | | | BK36/470/4/44 | 14.25 | 46.79 | 11.10 | 5.80 | 44.10 |
| | | | | | BK36/470/4/52 | 16.25 | 54.79 | 12.50 | 3.80 | 52.10 |
| | | | | | BK36/470/4/60 | 18.25 | 62.79 | 13.90 | 1.80 | 60.10 |
| | | | | | BK36/470/4/68 | 20.25 | 70.79 | 15.30 | – | – |
| BG 40H BS 100B Winch | 2800 | 2500 | 4000 | 24900 | BK36/470/3/30 | 13.25 | 33.16 | 7.75 | 9.45 | 30.90 |
| | | | | | BK36/470/3/36 | 15.25 | 39.16 | 8.65 | 7.45 | 36.90 |
| | | | | | BK36/470/3/45 | 18.25 | 48.16 | 10.00 | 4.45 | 45.90 |
| | | | | | BK36/470/4/40 | 13.25 | 42.79 | 10.30 | 9.45 | 40.50 |
| | | | | | BK36/470/4/48 | 15.25 | 50.79 | 11.80 | 7.45 | 48.50 |
| | | | | | BK36/470/4/56 | 17.25 | 58.79 | 13.20 | 5.45 | 56.50 |
| | | | | | BK36/470/4/64 | 19.25 | 66.79 | 14.60 | 3.45 | 64.50 |
| | | | | | BK36/470/4/72 | 21.25 | 74.79 | 16.00 | 1.45 | 72.50 |
| | | | | | BK36/470/4/80 | 23.25 | 82.79 | 17.50 | 0.45 | 80.50 |
| BG 40H BS 100B Cylinder | 3000 | 2700 | 4420 | 24900 | BK36/470/3/30 | 13.25 | 33.16 | 7.75 | 8.20 | 30.40 |
| | | | | | BK36/470/3/36 | 15.25 | 39.16 | 8.65 | 7.45 | 36.40 |
| | | | | | BK36/470/3/45 | 18.25 | 48.16 | 10.00 | 4.45 | 45.40 |
| | | | | | BK36/470/4/40 | 13.25 | 42.79 | 10.30 | 8.20 | 40.10 |
| | | | | | BK36/470/4/48 | 15.25 | 50.79 | 11.80 | 7.45 | 48.10 |
| | | | | | BK36/470/4/56 | 17.25 | 58.79 | 13.20 | 5.45 | 56.10 |
| | | | | | BK36/470/4/64 | 19.25 | 66.79 | 14.60 | 3.45 | 64.10 |
| | | | | | BK36/470/4/72 | 21.25 | 74.79 | 16.00 | 1.45 | 72.10 |
| | | | | | BK36/470/4/80 | 23.25 | 82.79 | 17.50 | – | – |

End of Table A.4.2

| 1 | 2 | 3 | 4 | 5 | 6 | 7 | 8 | 9 | 10 | 11 |
|---|---|---|---|---|---|---|---|---|---|---|
| BG 48 BS 160 Cylinder | 3000 | 2700 | 4802 | 33715 | BK48/559/4/40 | 13.50 | 44.37 | 13.80 | 8.50 | 42.00 |
| | | | | | BK48/559/4/48 | 15.50 | 52.37 | 16.50 | 8.50 | 50.00 |
| | | | | | BK48/559/4/56 | 17.50 | 60.37 | 19.20 | 8.50 | 58.00 |
| | | | | | BK48/559/4/64 | 19.50 | 68.37 | 21.90 | 8.50 | 66.00 |
| | | | | | BK48/559/4/80 | 23.75 | 85.17 | 24.50 | 7.76 | 82.70 |
| | | | | | BK36/559/5/70 | 17.25 | 73.00 | 17.80 | 8.50 | 70.50 |
| | | | | | BK36/559/5/80 | 19.25 | 83.00 | 20.10 | 8.50 | 80.50 |
| | | | | | BK36/559/5/100 | 23.75 | 105.00 | 24.50 | 7.76 | 102.20 |
| BG 48 BS 180 Cylinder | 3000 | 2700 | 4802 | 33715 | BK48/559/4/40 | 13.50 | 44.37 | 13.80 | 8.50 | 42.00 |
| | | | | | BK48/559/4/48 | 15.50 | 52.37 | 16.50 | 8.50 | 50.00 |
| | | | | | BK48/559/4/56 | 17.50 | 60.37 | 19.20 | 8.50 | 58.00 |
| | | | | | BK48/559/4/64 | 19.50 | 68.37 | 21.90 | 8.50 | 66.00 |
| | | | | | BK48/559/4/80 | 23.75 | 85.17 | 24.50 | 7.76 | 82.70 |
| | | | | | BK36/559/5/70 | 17.25 | 73.00 | 17.80 | 8.50 | 70.50 |
| | | | | | BK36/559/5/80 | 19.25 | 83.00 | 20.10 | 8.50 | 80.50 |
| | | | | | BK36/559/5/100 | 23.75 | 105.00 | 24.50 | 7.76 | 102.20 |

Notes: 1. The notation of the parameters is shown in Figure A.4.1. 2. The following abbreviations are taken in the table: Winch = drilling tool feed by means of winch; Cylinder = drilling tool feed by means of hydraulic winch. 3. Diameters of Bauer drilling tool (short augers, drill buckets or rotary core bits): 520, 600, 650, 700, 780, 800, 900, 1000, 1060, 1180, 1200, 1350, 1500, 1650, 1800, 1830, 2000, 2320 and 2500 mm. 4. Diameters of Bauer casing pipes (diameter of drilling tool is given in parenthesis): 620 (520), 750 (650), 880 (780), 1000 (900), 1180 (1060), 1300 (1180), 1500 (1350), 1650 (1500), 1800 (1650), 2000 (1830), 2200 (2000) and 2500 (2320) mm. 5. Example of Code designation example of Kelly bar of Model BK15 with outside diameter of 343 mm that includes 3 sections and intended for drilling of boreholes 18 m deep (rated): BK15/343/3/18.

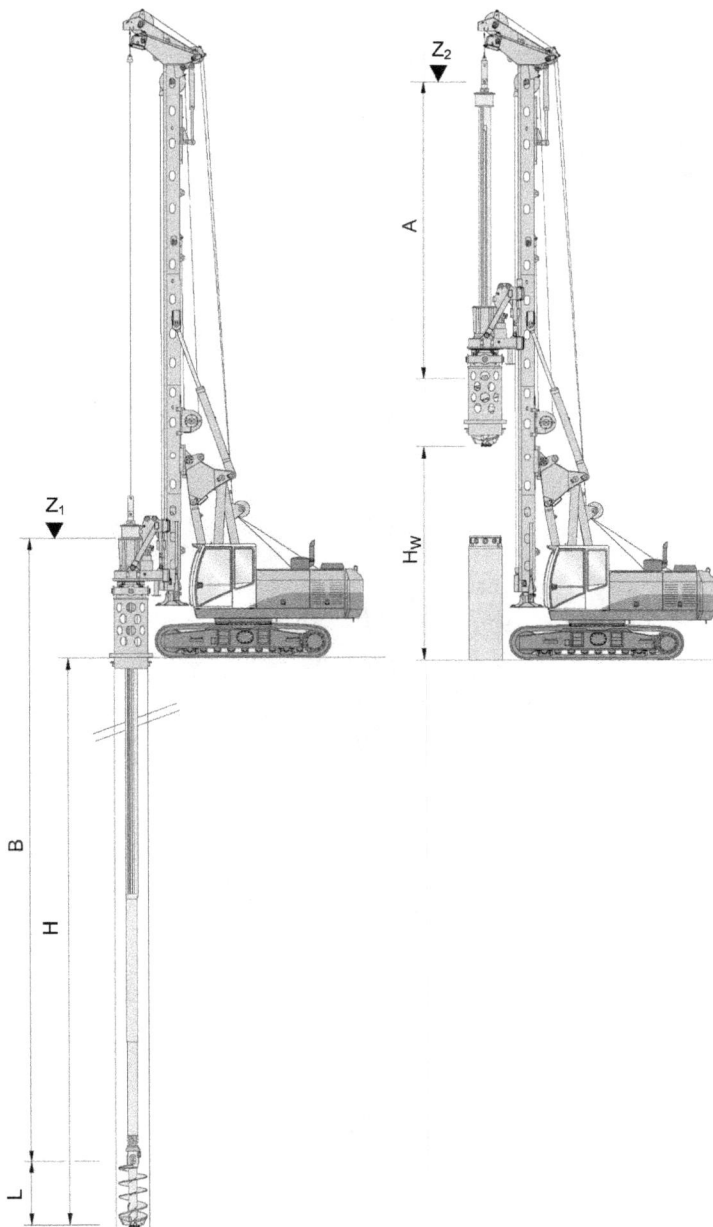

Figure A.4.1. Parameters of borehole drilling with casing pipe using short auger and Kelly bar: $B$ = length of Kelly bar; $L$ = length of drilling tool; $H$ = drilling depth

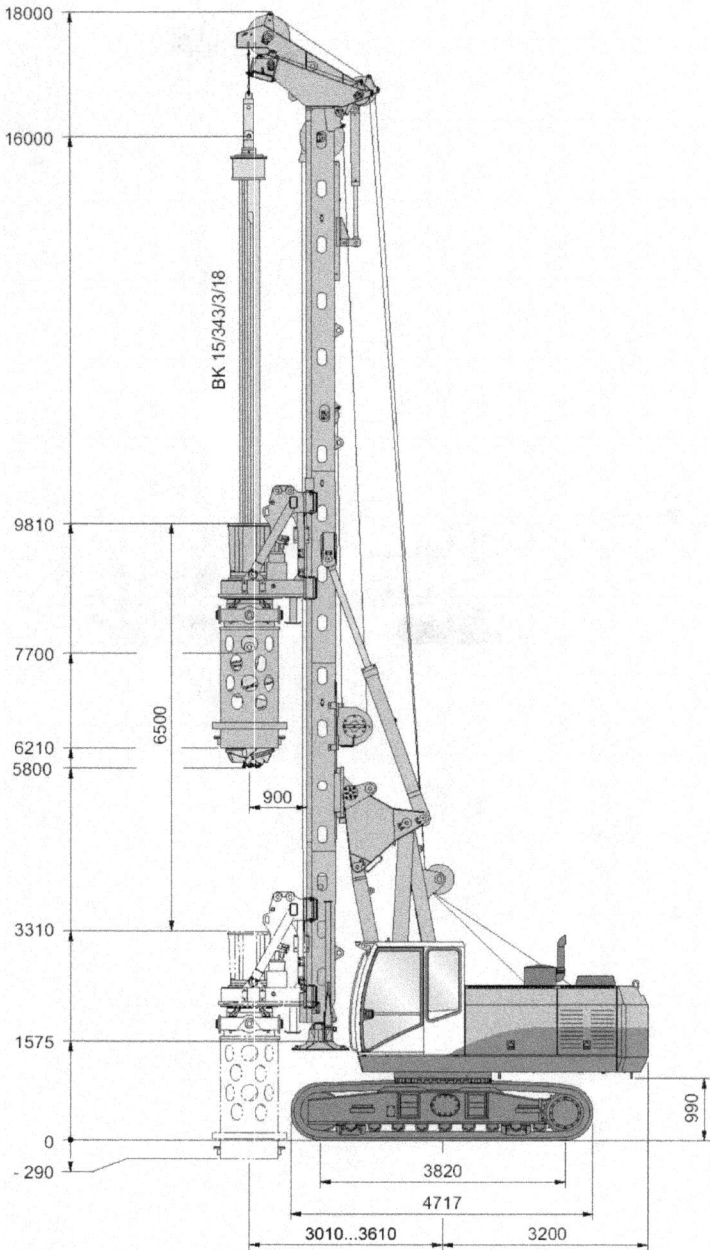

Figure A.4.2. Bauer BG 15H  BT 40 Rig with equipment to drill boreholes with casing pipe using short auger and Kelly bar (drilling tool feed by means of winch)

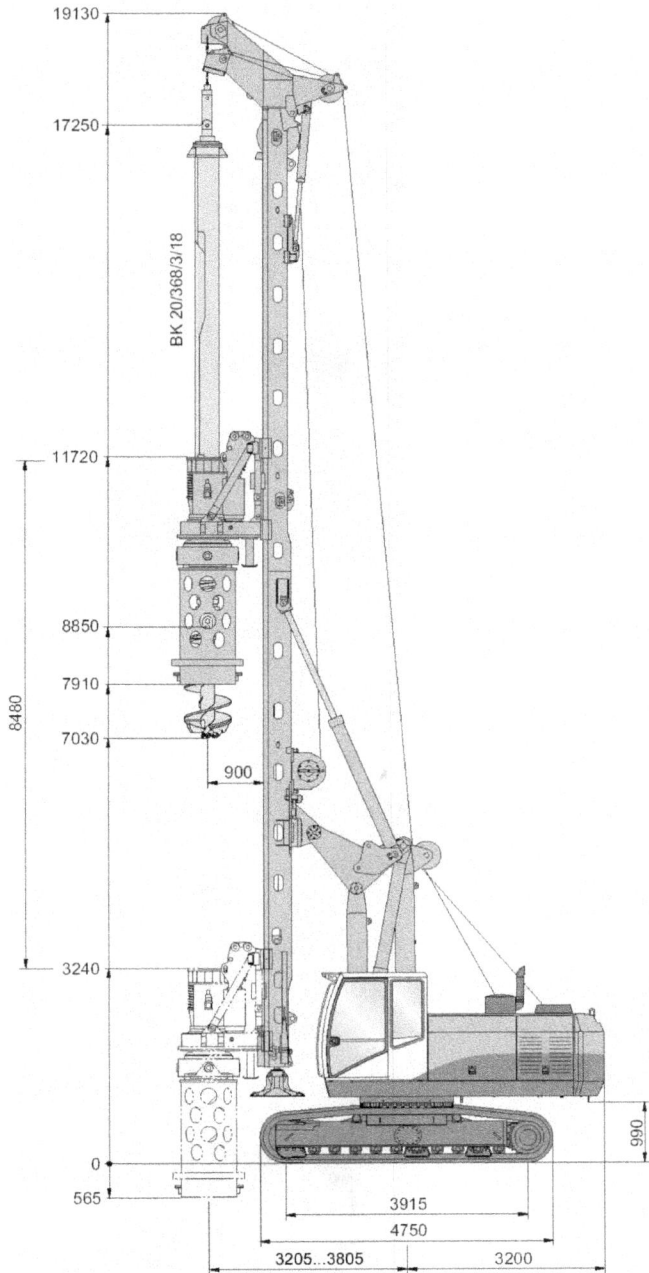

Figure A.4.3. Bauer BG 18H BT 50 Rig with equipment to drill boreholes with casing pipe using short auger and Kelly bar (drilling tool feed by means of winch)

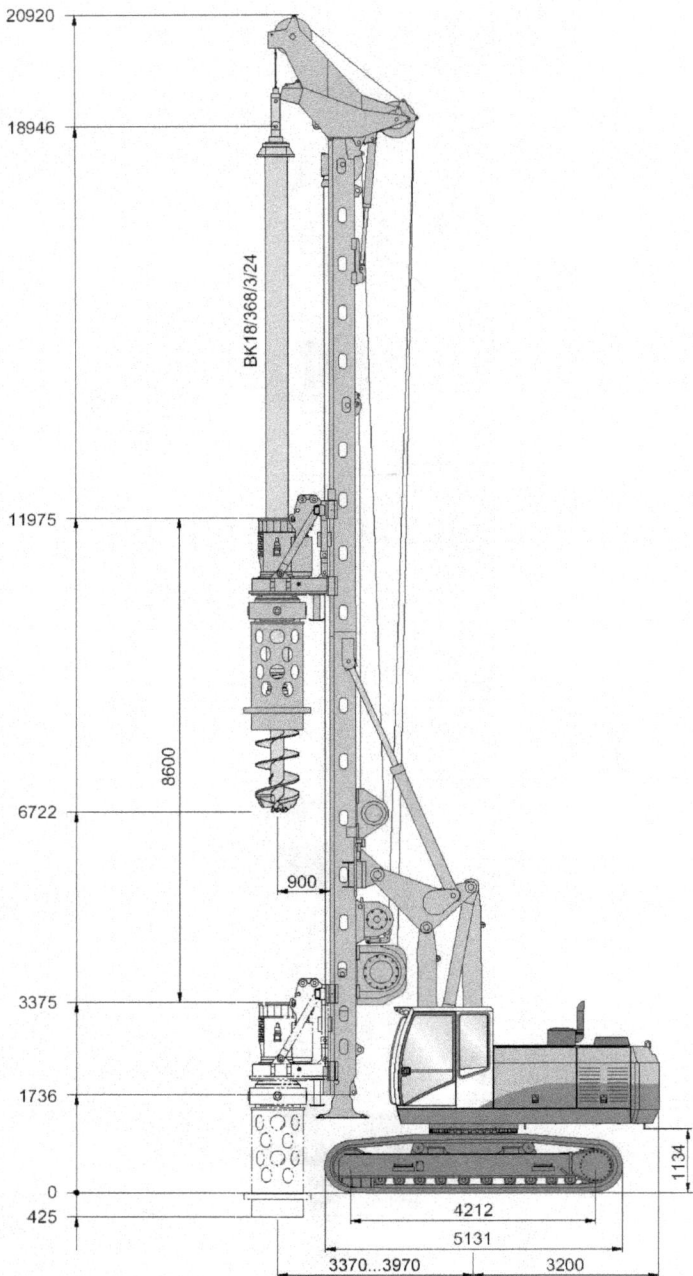

Figure A.4.4. Bauer BG 20H  BT 60 Rig with equipment to drill boreholes with casing pipe using short auger and Kelly bar (drilling tool feed by means of winch)

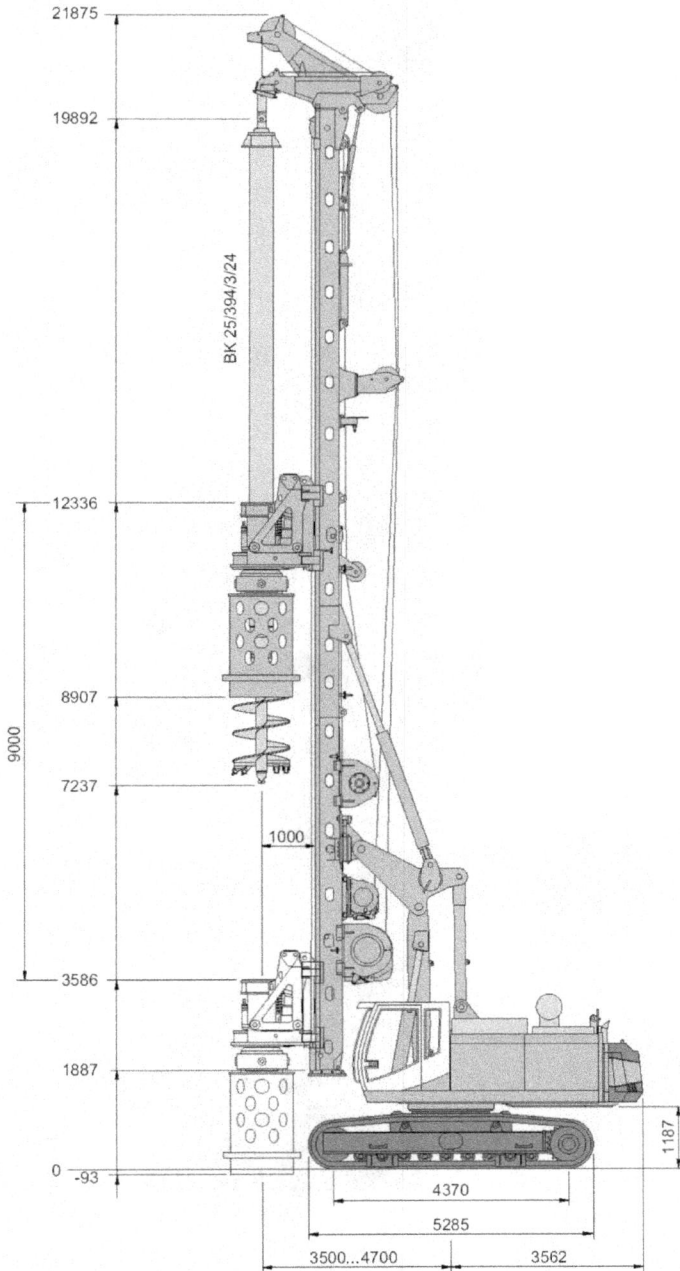

Figure A.4.5. Bauer BG 24H BS 70C Rig with equipment to drill boreholes with casing pipe using short auger and Kelly bar (drilling tool feed by means of winch)

151

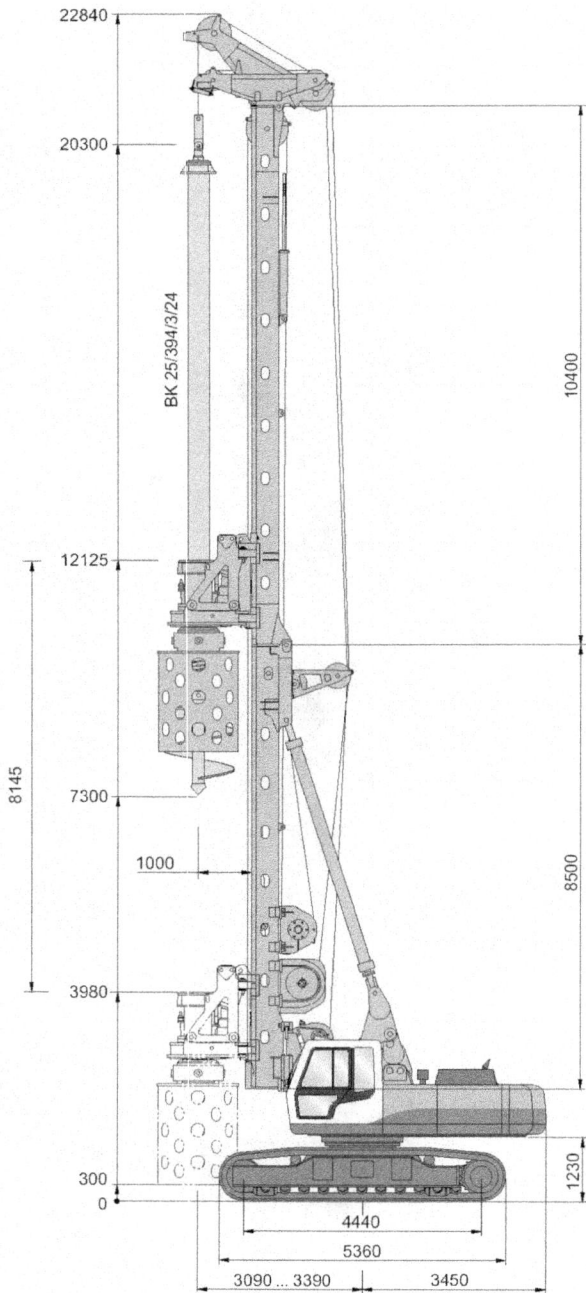

Figure A.4.6. Bauer BG 25C BH 60 Rig with equipment to drill boreholes with casing pipe using short auger and Kelly bar (drilling tool feed by means of winch)

152

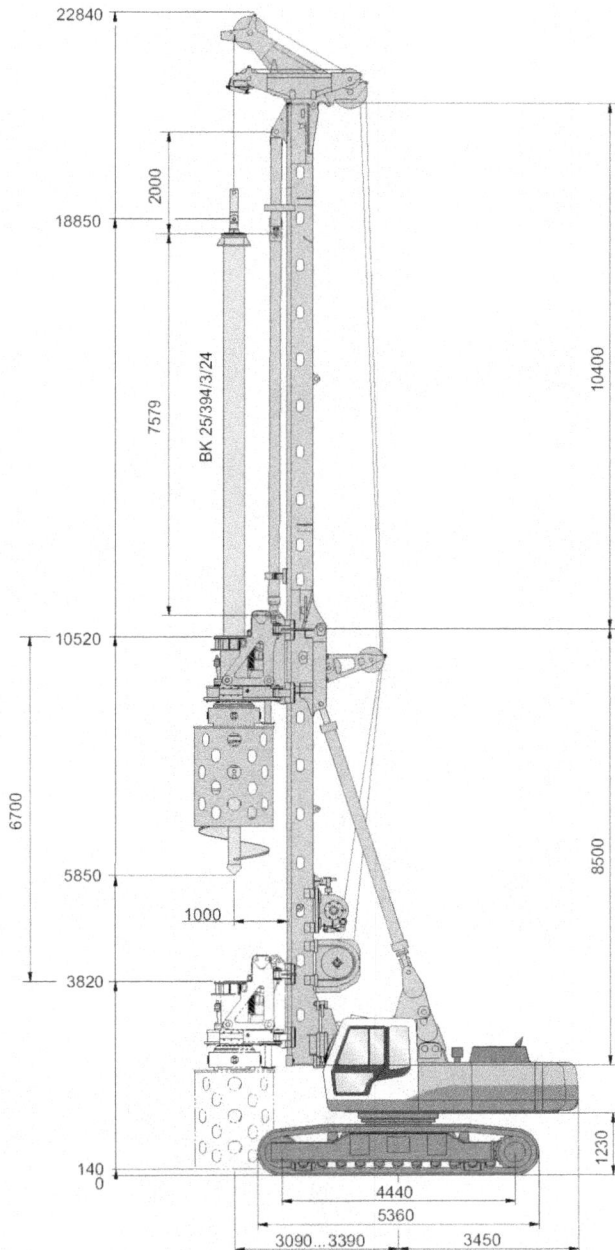

Figure A.4.7. Bauer BG 25C BH 60 Rig with equipment
to drill boreholes with casing pipe using short auger and Kelly bar
(drilling tool feed by means of hydraulic cylinder)

153

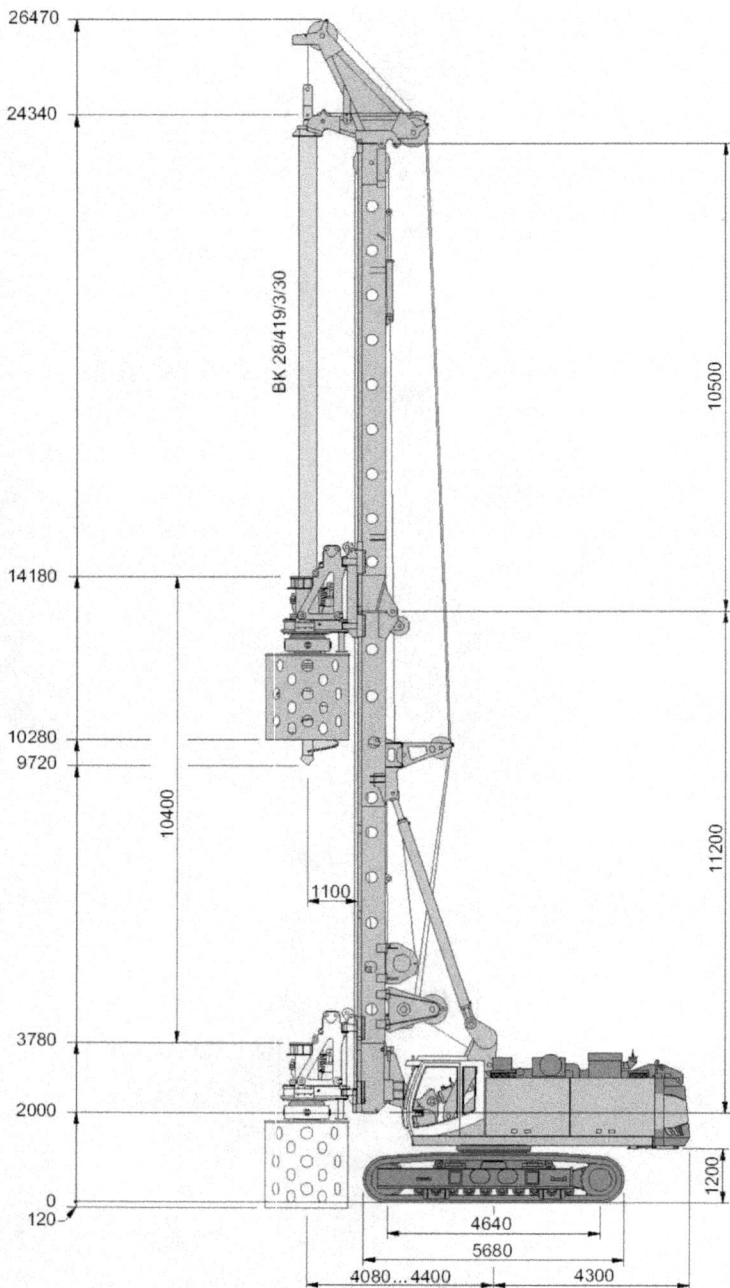

Figure A.4.8. Bauer BG 28  BS 80B Rig with equipment to drill boreholes with casing pipe using short auger and Kelly bar (drilling tool feed by means of winch)

154

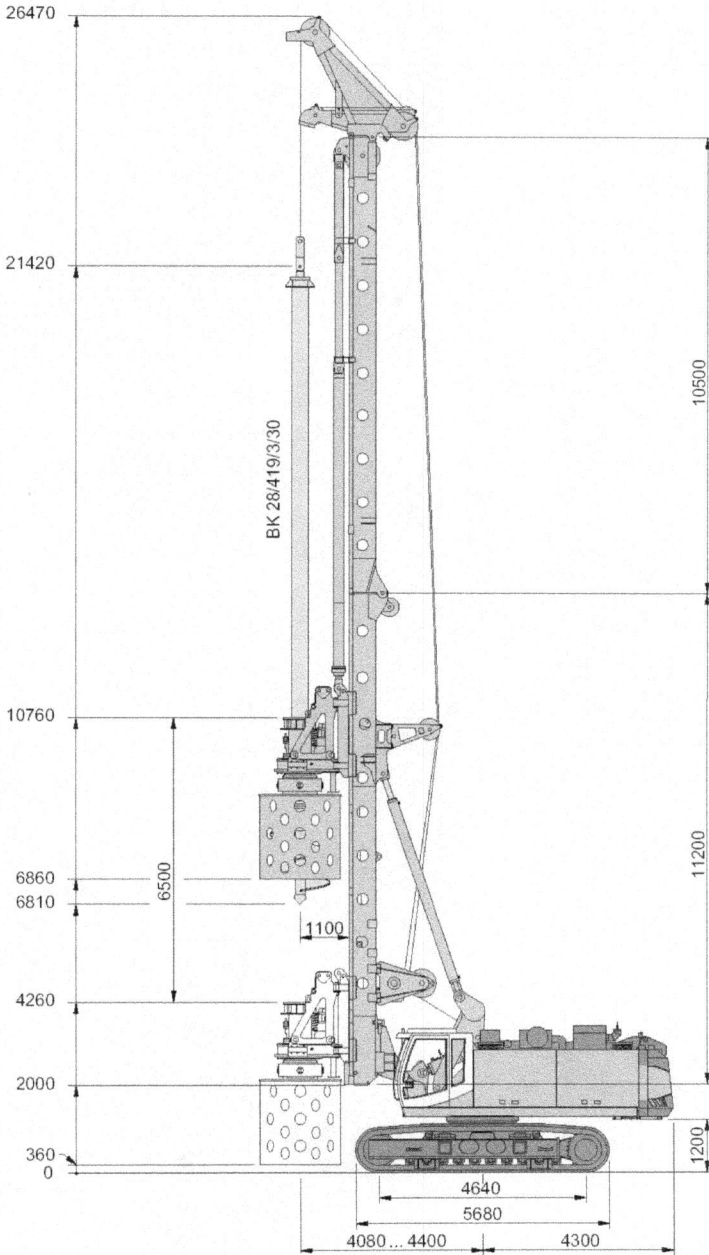

Figure A.4.9. Bauer BG 28  BS 80B Rig with equipment
to drill boreholes with casing pipe using short auger and Kelly bar
(drilling tool feed by means of hydraulic cylinder)

155

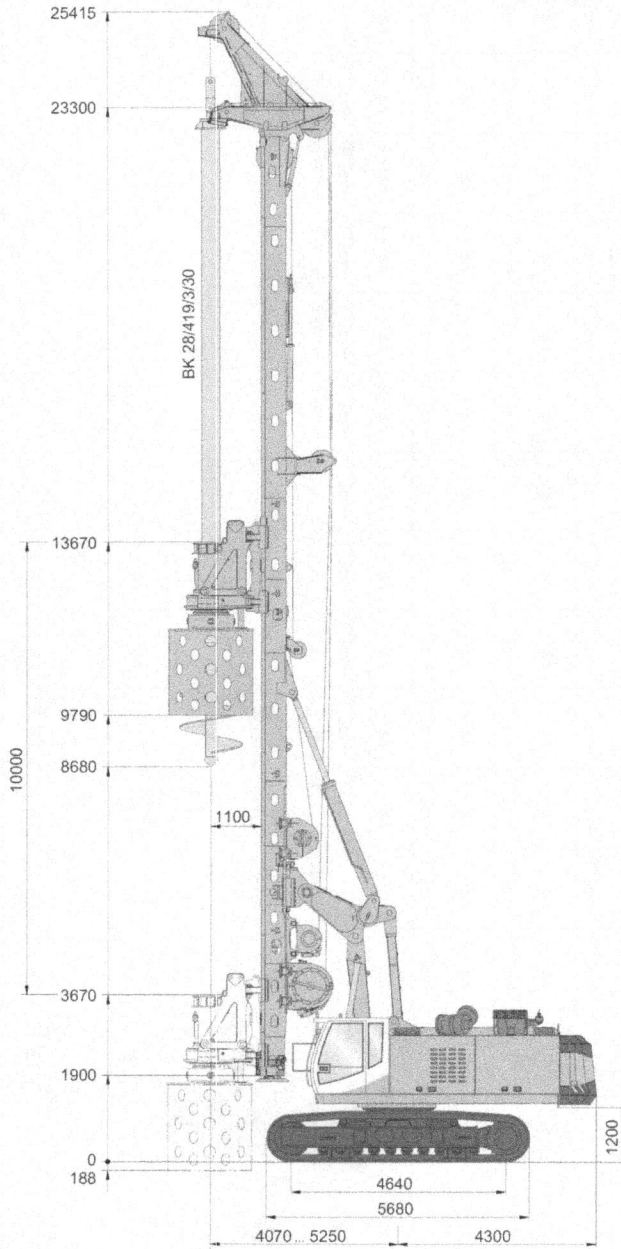

Figure A.4.10. Bauer BG 28H  BS 80B Rig with equipment to drill boreholes with casing pipe using short auger and Kelly bar (drilling tool feed by means of winch)

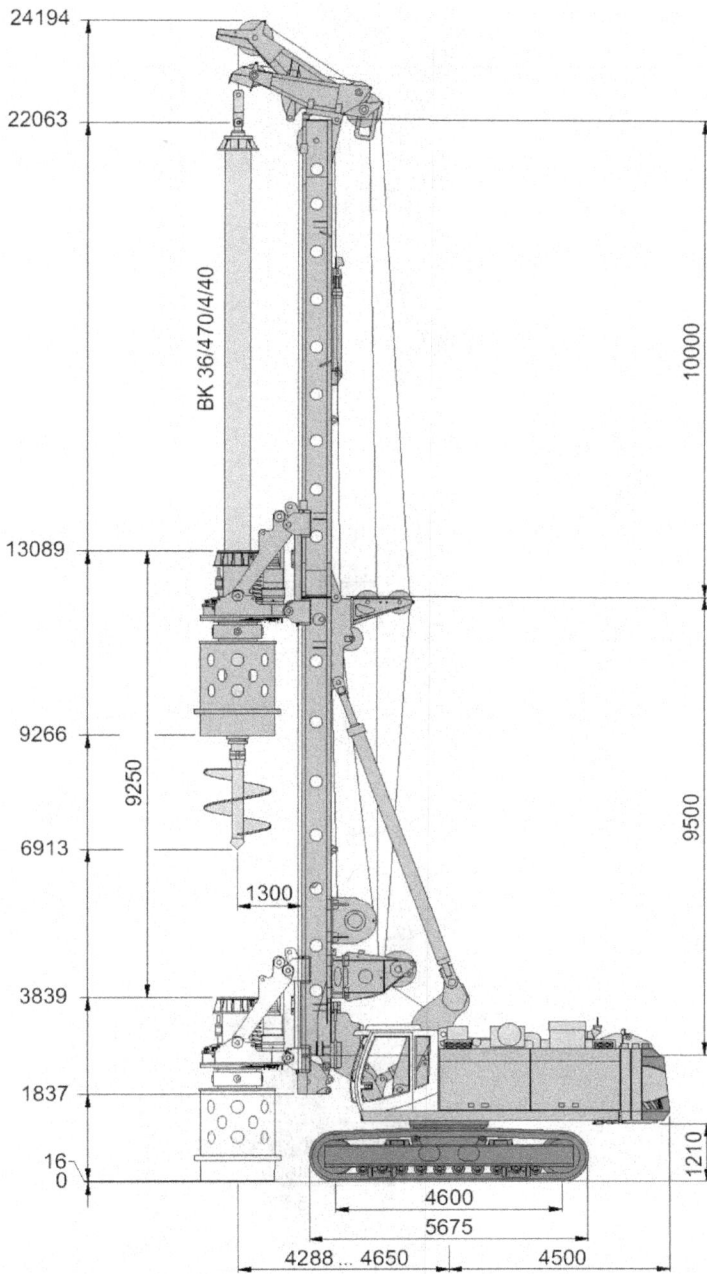

Figure A.4.11. Bauer BG 36  BS 80B Rig with equipment to drill boreholes with casing pipe using short auger and Kelly bar (drilling tool feed by means of winch)

157

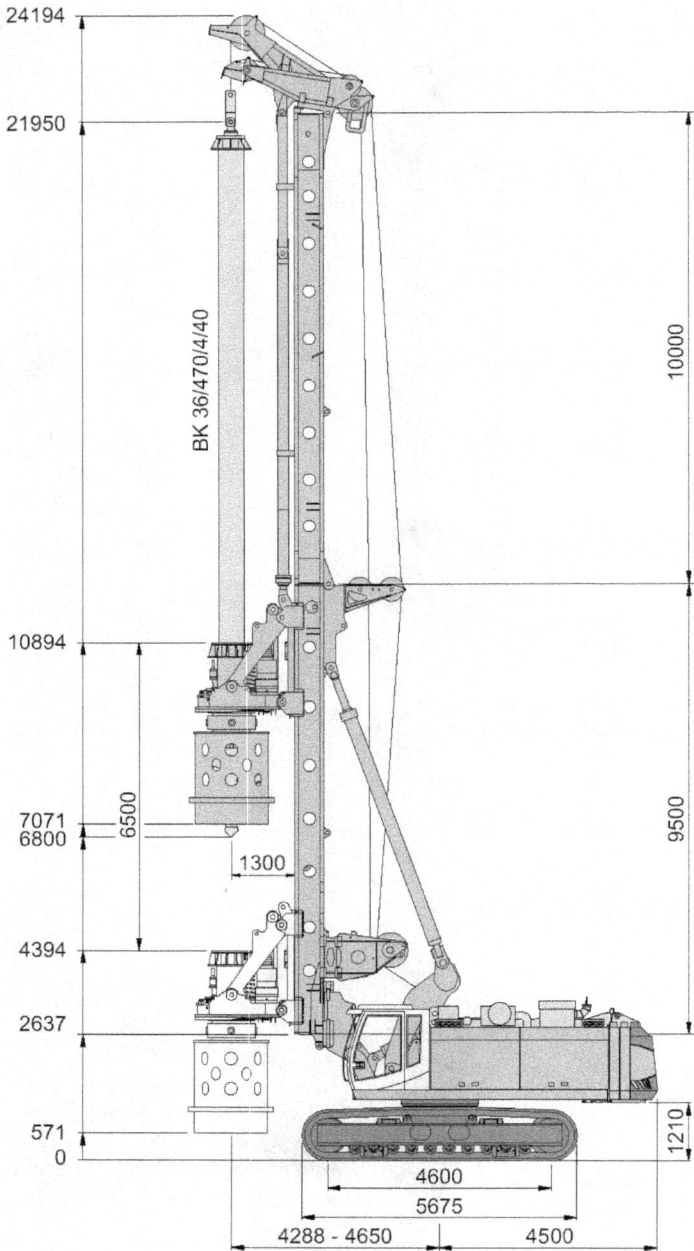

Figure A.4.12. Bauer BG 36  BS 80B Rig with equipment
to drill boreholes with casing pipe using short auger and Kelly bar
(drilling tool feed by means of hydraulic cylinder)

158

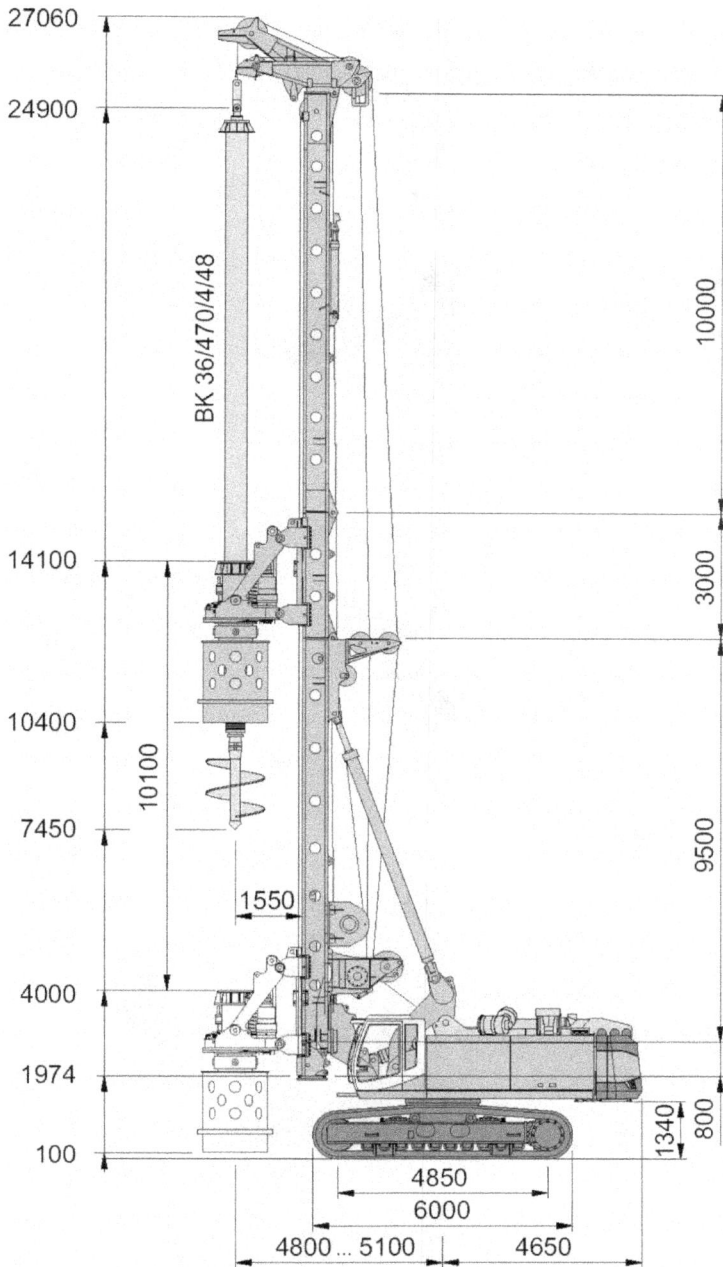

Figure A.4.13. Bauer BG 40  BS 100B Rig with equipment to drill boreholes with casing pipe using short auger and Kelly bar (drilling tool feed by means of winch)

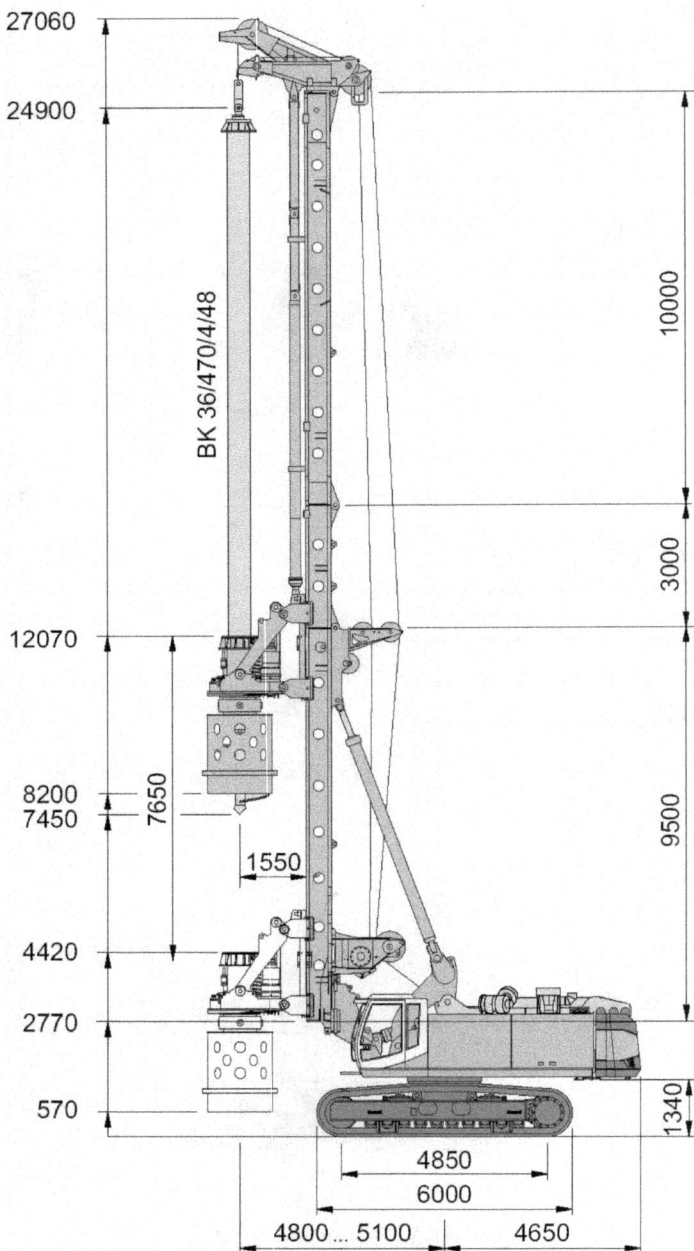

Figure A.4.14. Bauer BG 40  BS 100B Rig with equipment
to drill boreholes with casing pipe using short auger and Kelly bar
(drilling tool feed by means of hydraulic cylinder)

160

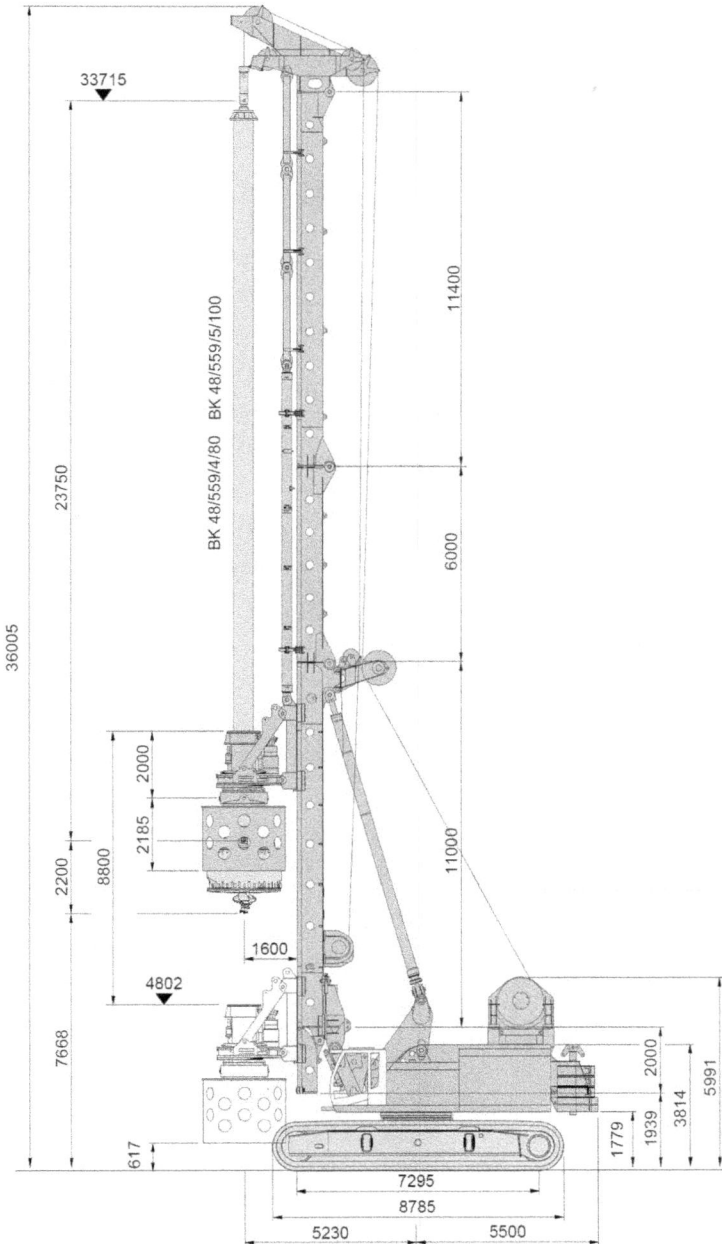

Figure A.4.15. Bauer BG 48  BS 160 Rig with equipment
to drill boreholes with casing pipe using short auger and Kelly bar
(drilling tool feed by means of hydraulic cylinder)

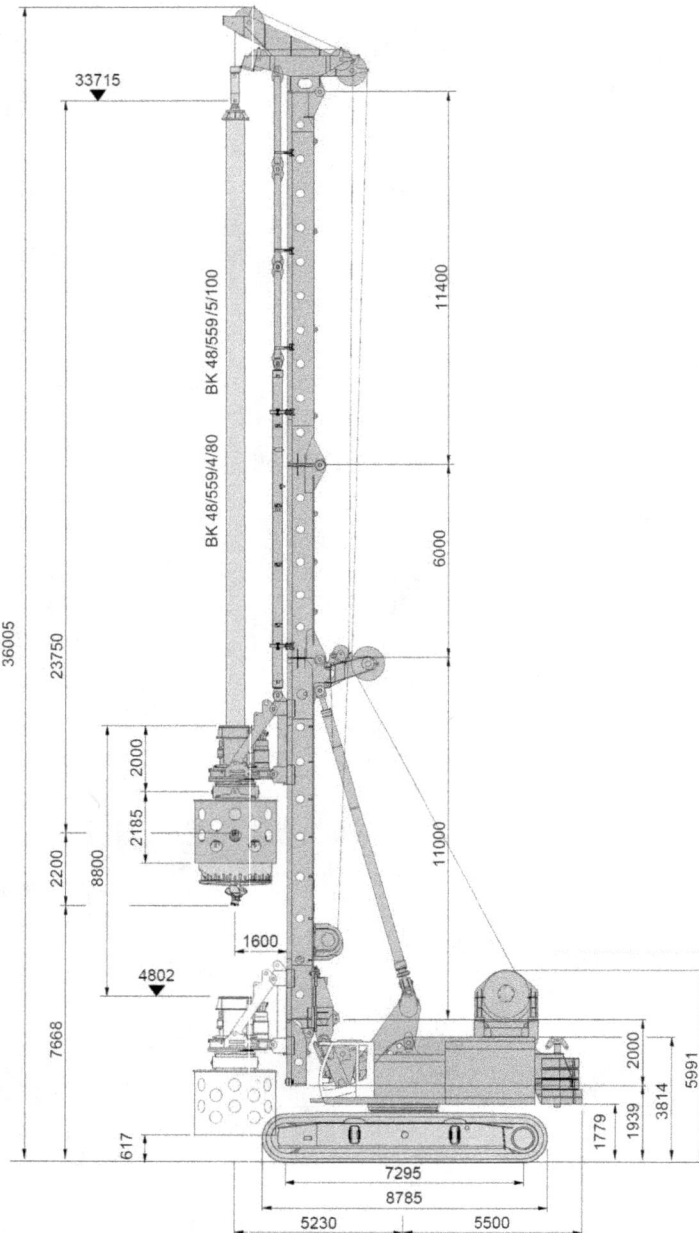

Figure A.4.16. Bauer BG 48  BS 180 Rig with equipment to drill boreholes with casing pipe using short auger and Kelly bar (drilling tool feed by means of hydraulic cylinder)

## Parameters of borehole drilling with continuous flight auger

| Model of Bauer rig | Method of drilling tool feed | Maximum drilling diameter, mm | Upfeed force of drilling tool, kN | Downfeed force of drilling tool, kN | Standard drilling | | Drilling with auger extension | | |
|---|---|---|---|---|---|---|---|---|---|
| | | | | | Depth, m | Auger length, m | Extension using Kelly bar, m | Depth, m | Auger length, m |
| 1 | 2 | 3 | 4 | 5 | 6 | 7 | 8 | 9 | 10 |
| BG 15H BT 40 | Jacking winch | 780 | 200 | 100 + auger weight | 11.3/12.3 | 13.4/13.4 | 6.0 | 17.3/18.3 | 13.4/13.4 |
| | Jacking and main winches | 780 | 420 | 100 + auger weight | 10.5/11.5 | 12.6/12.6 | 6.0 | 16.98/17.98 | 13.1/13.1 |
| BG 18H BT 50 | Jacking winch | 780 | 200 | 165 + auger weight | 12.43/13.43 | 14.24/14.24 | 6.0 | 18.43/19.43 | 14.24/14.24 |
| | Jacking and main winches | 780 | 480 | 165 + auger weight | 11.68/12.68 | 12.49/13.49 | 6.0 | 18.13/19.13 | 13.94/13.94 |
| BG 20H BT 60 | Jacking winch | 880 | 210 | 205 + auger weight | 13.45/14.65 | 15.6/15.6 | 6.0 | 19.4/20.6 | 15.6/15.6 |
| | Jacking and main winches | 880 | 550 | 205 + auger weight | 12.65/13.85 | 14.8/14.8 | 6.0 | 18.9/20.1 | 15.1/15.1 |
| BG 24H BS 70C | Jacking winch | 1000 | 330 | 200 + auger weight | 13.9/14.9 | 16.2/16.2 | 6.0 | 19.9/20.9 | 15.8/15.8 |
| | Jacking and main winches | 1000 | 730 | 200 + auger weight | 13.9/14.9 | 16.2/16.2 | 6.0 | 19.9/20.9 | 15.8/15.8 |
| BG 25C BH 60 | Jacking winch | 800 | 260 | 260 + auger weight | 14.2/15.2 | 17.25/17.25 | 6.0 | 20.2/21.2 | 17.25/17.25 |
| | Jacking hydraulic cyinder | 800 | 400 | Weight of auger rotation head + auger weight | 14.2/15.2 | 17.2/17.2 | 6.0 | 20.2/21.2 | 17.2/17.2 |

| 1 | 2 | 3 | 4 | 5 | 6 | 7 | 8 | 9 | 10 |
|---|---|---|---|---|---|---|---|---|---|
| BG 28  BS 80B | Jacking winch | 1200 | 330 | 200 + dead weight | 17.5 / 18.2 | 20.0 / 20.0 | 8.0 | 25.5 / 26.2 | 20.0 / 20.0 |
| | Jacking and main winches | 1200 | 830 | 200 + dead weight | 17.5 / 18.2 | 20.0 / 20.0 | 8.0 | 25.5 / 26.2 | 20.0 / 20.0 |
| | Jacking hydraulic cylinder | 1200 | 500 | Weight of auger rotation head + auger weight | 16.7 / 17.2 | 19.2 / 19.2 | 8.0 | 24.7 / 25.2 | 19.2 / 19.2 |
| BG 36  BS 80B | Jacking hydraulic cylinder | 1200 | 500 | Weight of auger rotation head + auger weight | 14.4 / 15.4 | 16.6 / 16.6 | | 20.2 / 21.4 | 16.6 / 16.6 |
| | Jacking winch | 1200 | 400 | 250 + auger weight | 14.8 / 16.0 | 17.2 / 17.2 | 6.0 | 20.8 / 22.0 | 17.2 / 17.2 |
| | Jacking hydraulic cylinder | 1200 | 600 | Weight of auger rotation head + auger weight | 17.4 / 18.6 | 20.3 / 20.3 | 6.0 | 23.6 / 24.8 | 20.3 / 20.3 |
| BG 40  BS 100B | Jacking winch | 1200 | 460 | 250 + auger weight | 18.0 / 24.0 | 20.9 / 20.9 | 6.0 | 19.2 / 25.2 | 20.9 / 20.9 |
| | Jacking and main winches | 1200 | 1060 | 250 + auger weight | 18.0 / 24.0 | 20.9 / 20.9 | 6.0 | 19.2 / 25.2 | 20.9 / 20.9 |

*Notes:* 1. In the numerator, there are indicated the drilling parameters when an auger cleaner is used; in the denominator, there are indicated the drilling parameters when no cleaner is used. 2. The diameters of the Bauer continuous augers (theoretical diameters of boreholes): 270, 370, 400, 550, 630, 750, 880 and 1200 mm. 3. Diameters of boreholes drilled using the double-rotary technology: 305, 406, 510, 620, 750, 880 and 1180 mm.

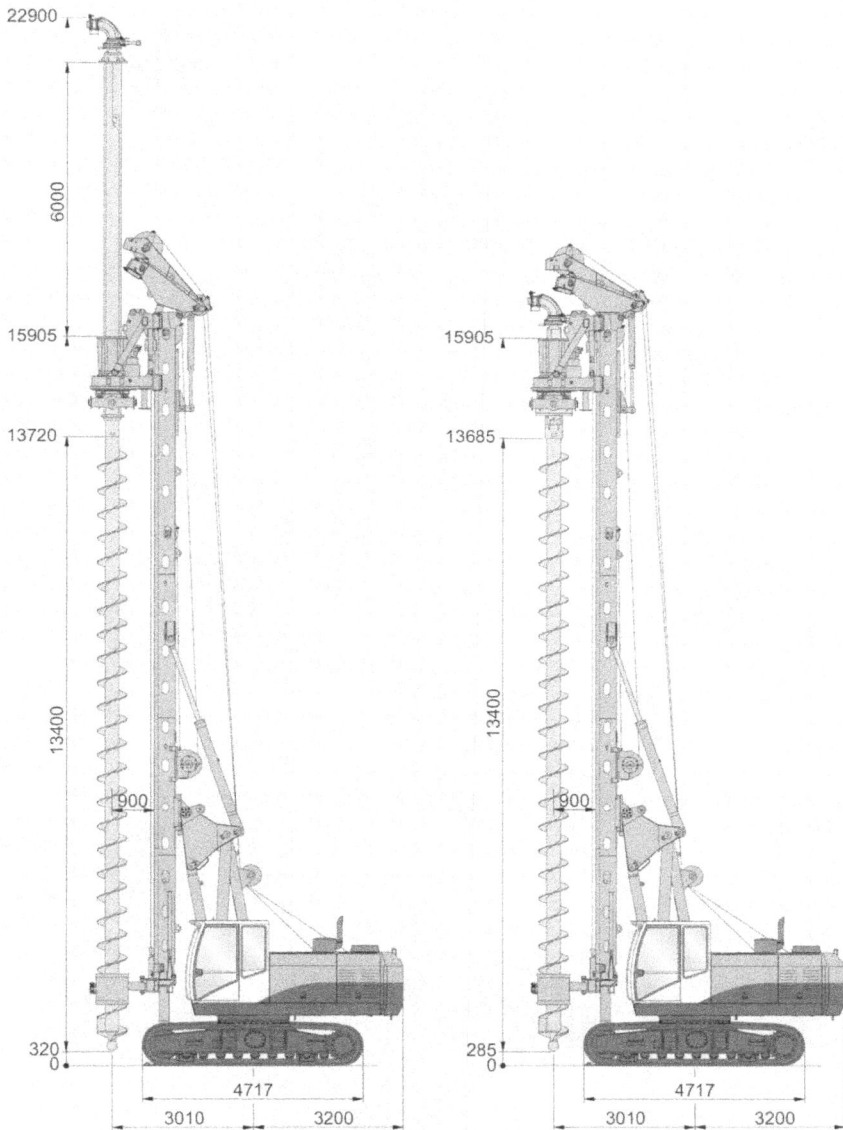

Figure A.4.17. Bauer BG 15H  BT 40 rig with equipment
to drill boreholes with continuous flight auger

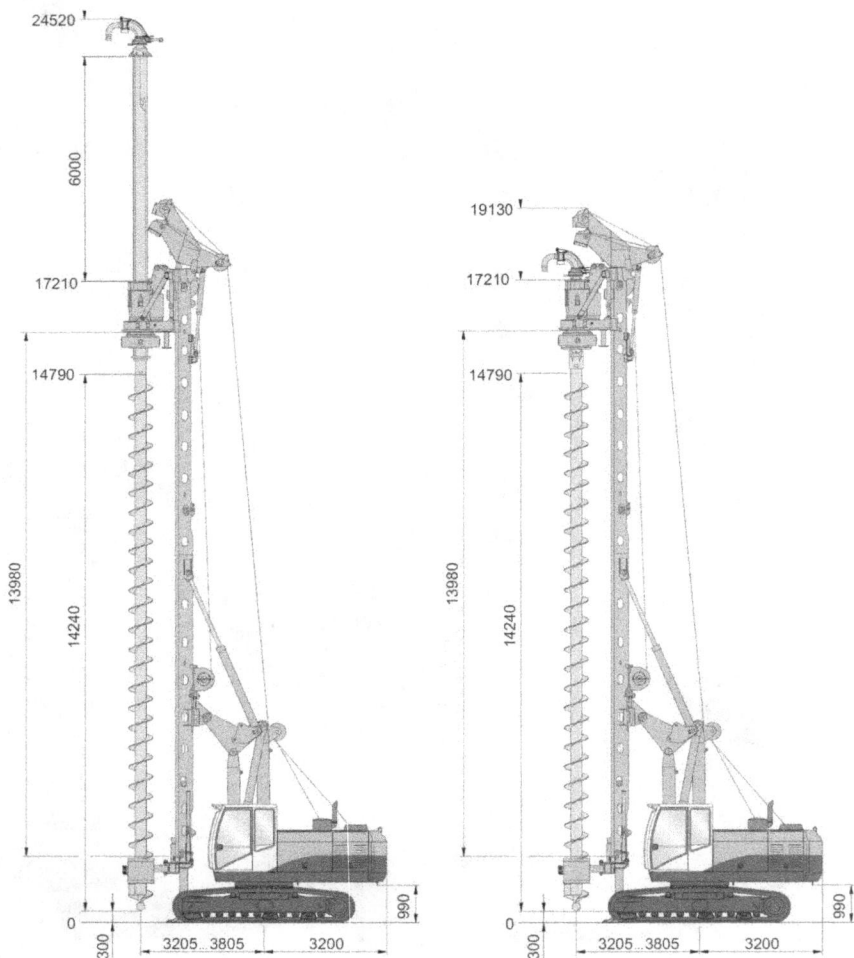

Figure A.4.18. Bauer BG 18H  BT 50 rig with equipment
to drill boreholes with continuous flight auger

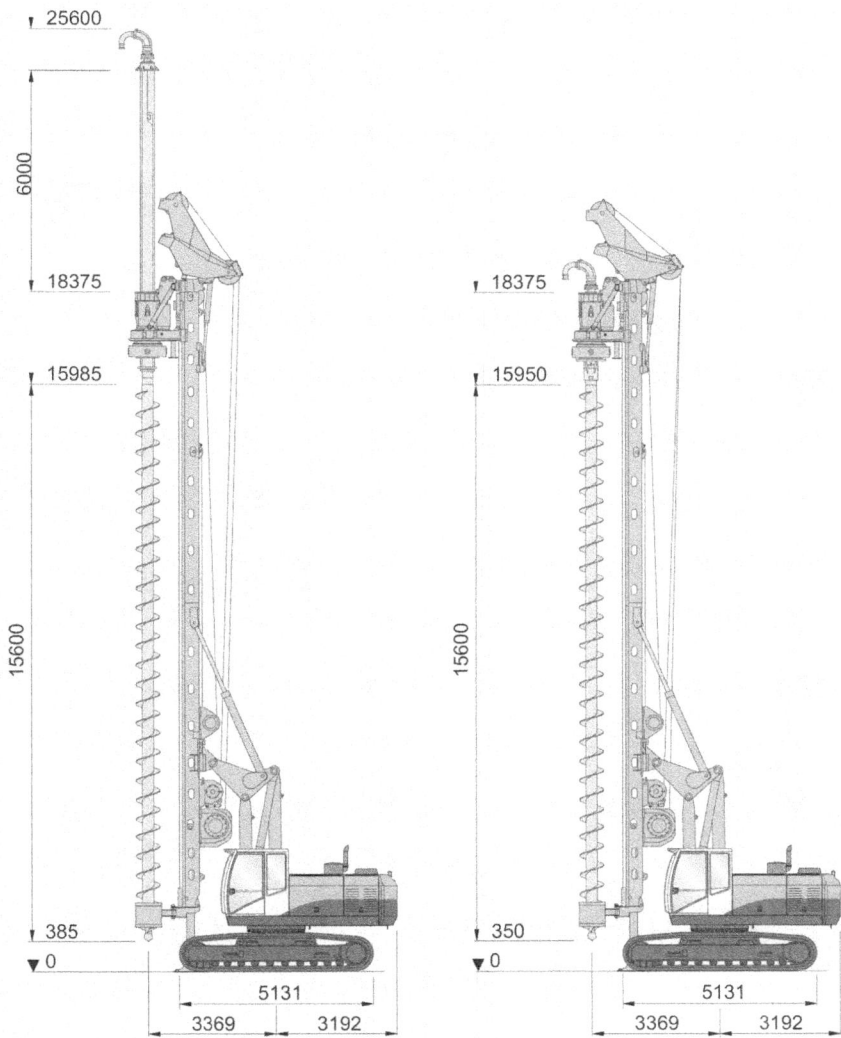

Figure A.4.19. Bauer BG 20H  BT 60 rig with equipment
to drill boreholes with continuous flight auger

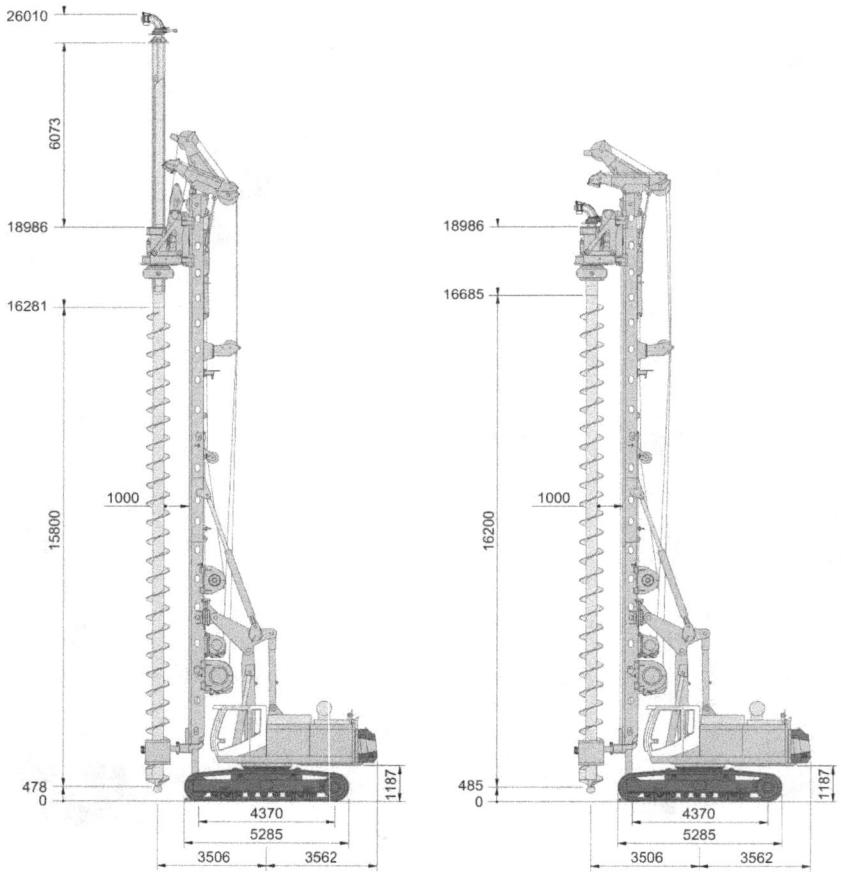

Figure A.4.20. Bauer BG 24H  BS 70C rig with equipment
to drill boreholes with continuous flight auger

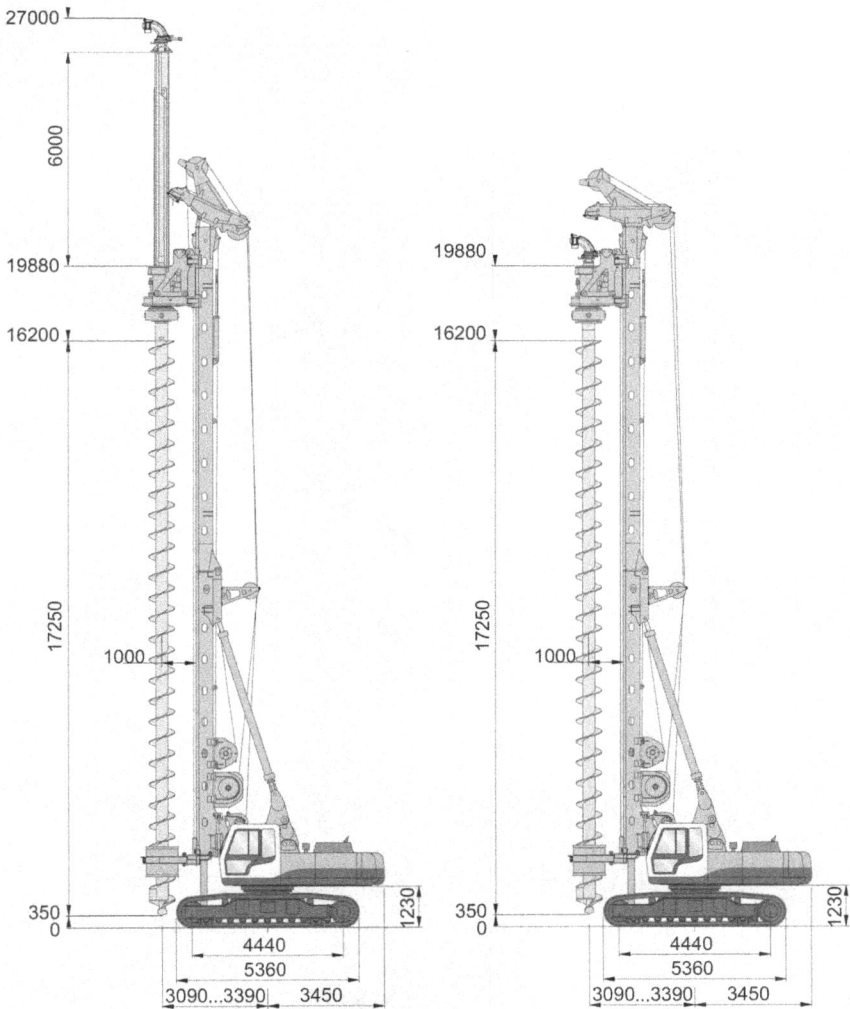

Figure A.4.21. Bauer BG 25C  BH 60 rig with equipment
to drill boreholes with continuous flight auger

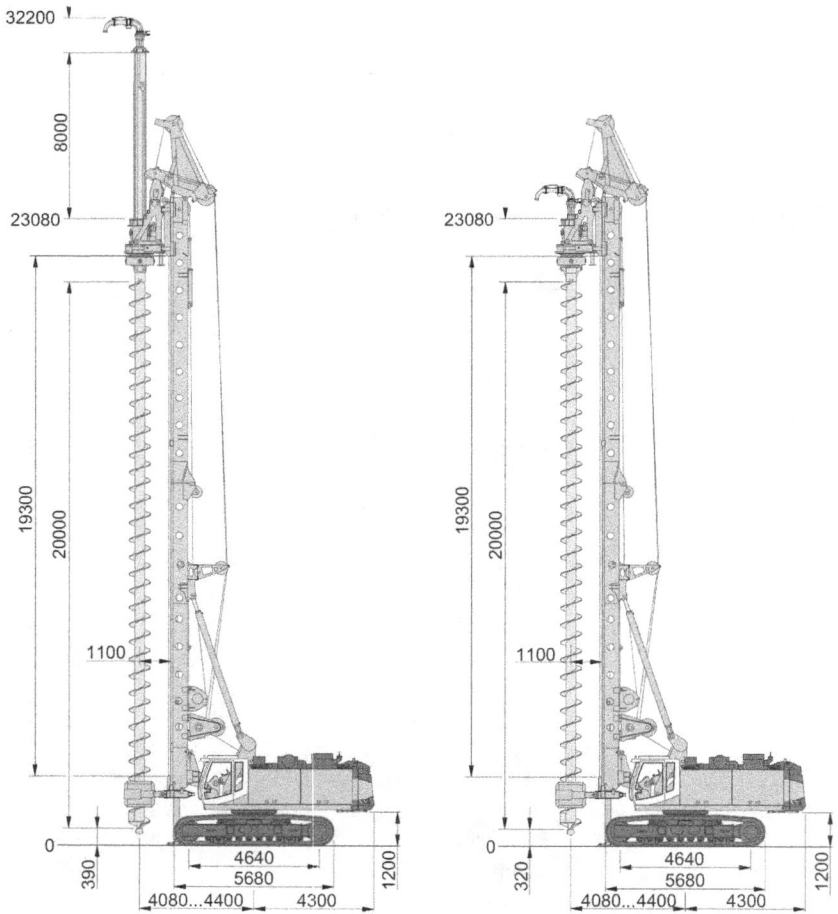

Figure A.4.22. Bauer BG 28  BS 80B rig with equipment
to drill boreholes with continuous flight auger

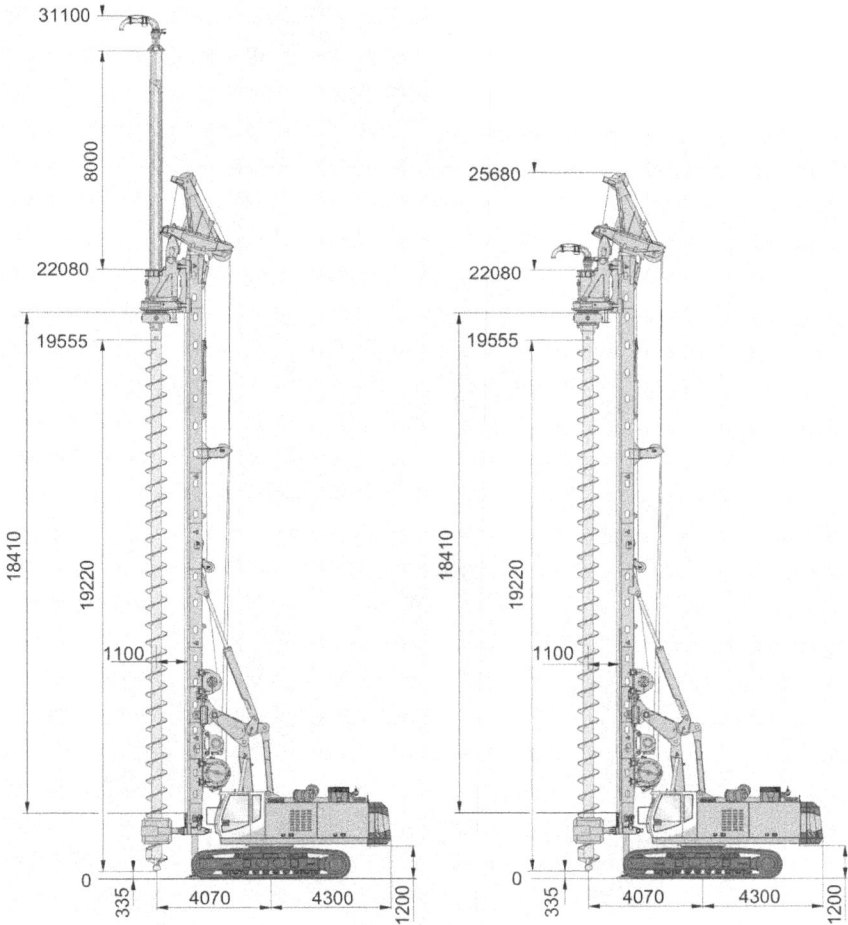

Figure A.4.23. Bauer BG 28H  BS 80B rig with equipment
to drill boreholes with continuous flight auger

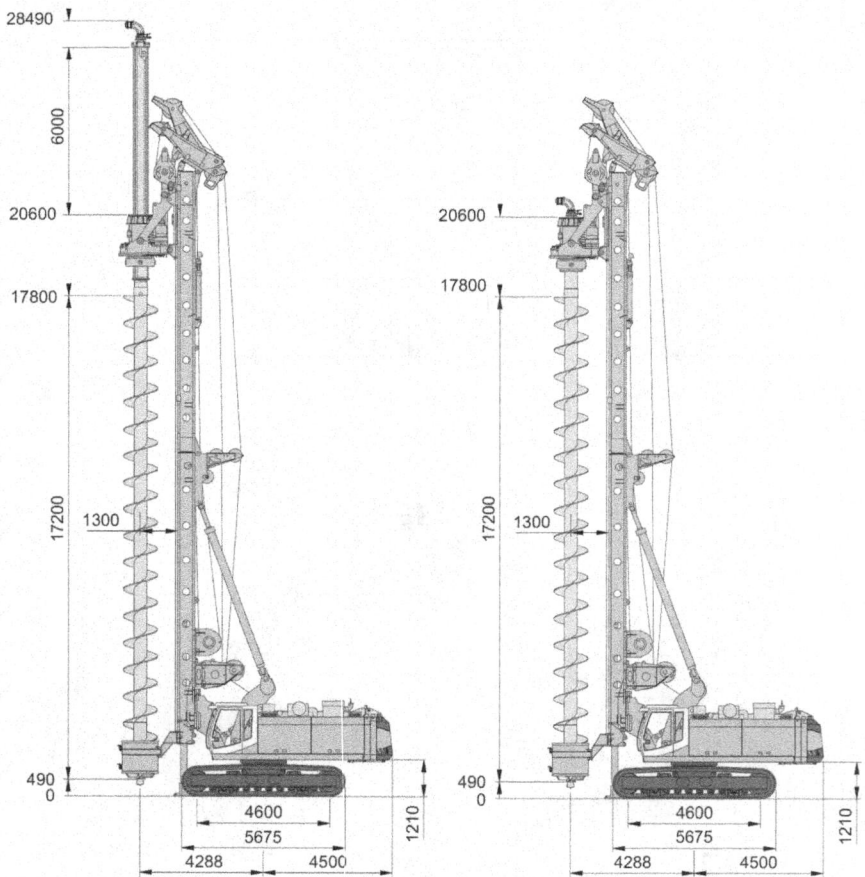

Figure A.4.24. Bauer BG 36 BS 80B rig with equipment
to drill boreholes with continuous flight auger

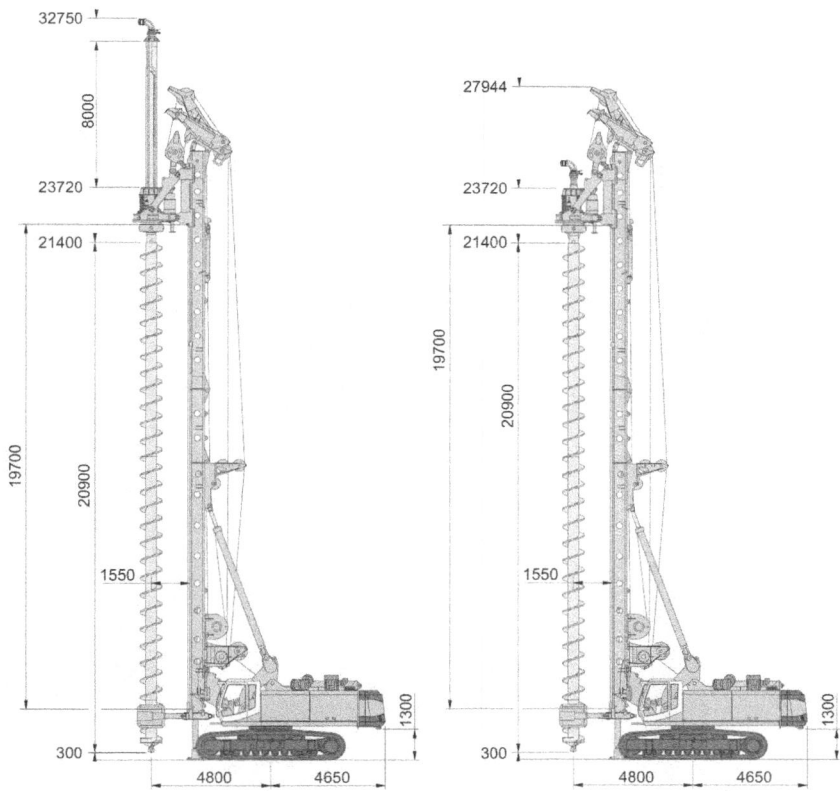

Figure A.4.25. Bauer BG 40  BS 100 rig with equipment
to drill boreholes with continuous flight auger

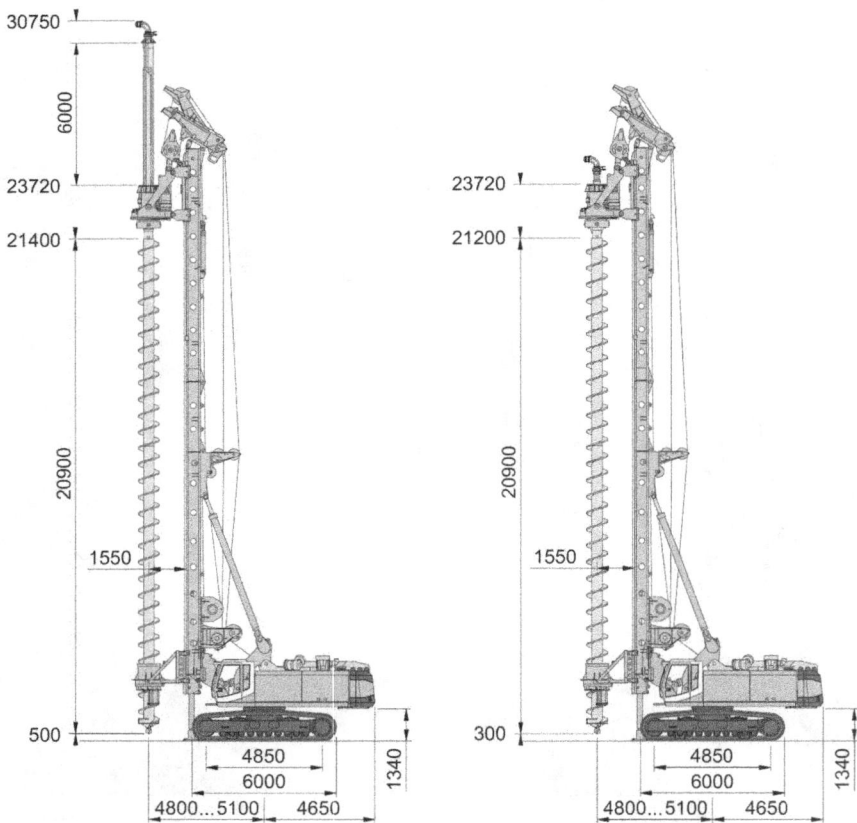

Figure A.4.26. Bauer BG 40 BS 100B rig with equipment
to drill boreholes with continuous flight auger

*Table A.4.4*

## Borehole-reaming depth, m, by means of RTG Rammtechnik helical tools

| $D_1$, mm | 360 | | | 440 | | | 440 | | | 510 | | | 510 | | | 620 | | |
|---|---|---|---|---|---|---|---|---|---|---|---|---|---|---|---|---|---|---|
| $D_2$, mm | 254 | | | 254 | | | 368 | | | 254 | | | 368 | | | 368 | | |
| $L$, mm | 3025 | | | 3025 | | | 3400 | | | 3025 | | | 3400 | | | 3400 | | |
| Bauer rig model | STM | K | KLM | STM | K | KLM | STM | K | KLM | STM | K | KLM | STM | K | KLM | STM | K | KLM |
| BG 12H | 10.5 | 14.5 | – | – | – | – | – | – | – | – | – | – | – | – | – | – | – | – |
| BG 15H | 11.5 | 17.5 | – | – | – | – | – | – | – | – | – | – | – | – | – | – | – | – |
| BG 18H | – | – | – | 13.0 | 21.0 | – | 13.0 | 21.0 | – | 13.0 | 21.0 | – | 13.0 | 21.0 | – | – | – | – |
| BG 20 | – | – | – | 13.0 | 21.0 | – | 13.0 | 21.0 | – | 13.0 | 21.0 | – | 13.0 | 21.0 | – | – | – | – |
| BG 20H | – | – | – | 14.0 | 22.0 | – | 14.0 | 22.0 | – | 14.0 | 22.0 | – | 14.0 | 22.0 | – | – | – | – |
| BG 22H | – | – | – | – | – | – | 15.0 | 23.0 | 25.5 | – | – | – | 15.0 | 23.0 | 25.5 | – | – | – |
| BG 24 | – | – | – | – | – | – | 16.0 | 24.0 | 26.5 | – | – | – | 16.0 | 24.0 | 26.5 | 16.0 | 24.0 | 26.5 |
| BG 24H | – | – | – | – | – | – | 14.5 | 22.5 | 25.0 | – | – | – | 14.5 | 22.5 | 25.0 | 14.5 | 22.5 | 25.0 |
| BG28 | – | – | – | – | – | – | – | – | – | – | – | – | 18.0 | 26.0 | 28.5 | 18.0 | 26.0 | 28.5 |
| BG 28H | – | – | – | – | – | – | – | – | – | – | – | – | 17.5 | 25.5 | 28.0 | 17.5 | 25.5 | 28.0 |
| BG 36 | – | – | – | – | – | – | – | – | – | – | – | – | 17.5 | 25.5 | 29.5 | 17.5 | 25.5 | 29.5 |
| BG 36H | – | – | – | – | – | – | – | – | – | – | – | – | 17.0 | 25.0 | 29.0 | 17.0 | 25.0 | 29.0 |
| BG 40 | – | – | – | – | – | – | – | – | – | – | – | – | 18.5 | 26.5 | 30.5 | 18.5 | 26.5 | 30.5 |
| RG 20S | – | – | – | 20.0 | 24.0 | – | 20.0 | 24.0 | – | 20.0 | 24.0 | – | 20.0 | 24.0 | – | – | – | – |
| RG 25S | – | – | – | 23.0 | 27.0 | – | 23.0 | 27.0 | – | 23.0 | 27.0 | – | 23.0 | 27.0 | – | 23.0 | 27.0 | – |

*Note.* The following abbreviations are taken in the table: $D_1$ = reamer diameter (theoretical diameter of borehole); $D_2$ = diameter of drill pipe; $L$ = reamer length; STM = standard mast of drilling rig; K = extension by means of Kelly bar; KLM = extension by means of Kelly bar and lattice column mast.

# TECHNICAL FEATURES OF SOILMEC RIGS AND EQUIPMENT

Table A.5.1

**Technical features of SoilMec CM rigs**
**designed for drilling boreholes with continuous flight auger**

| Feature | Rig model | | | | | |
|---|---|---|---|---|---|---|
| | CM-50 | CM-70 | CM-700 | CM-120 | CM-1200 | |
| 1 | 2 | 3 | 4 | 5 | 6 | |
| Maximum drilling diameter, mm | 900 900 | 1000 1000 | 1000 1000 | 1200 1400 | 1200 1400 | |
| Maximum drilling depth without auger extension, m | 17.5 19.0 | 21.0 22.0 | 21.5 23.0 | 23.0 24.5 | 26.0 27.5 | |
| Maximum drilling depth with auger extension, m | 23.5 25.0 | 27.0 28.0 | 27.5 29.0 | 29.0 30.5 | 32.0 33.5 | |
| Maximum torsion torque, kN·m | 86 | 156 | 172 | 300 | 300 | |
| Auger rotation speed in drilling, rpm | 41 | 25 | 25 | 25 | 25 | |
| Auger pulling force, kN | 510 | 680 | 680 | 1160 | 1160 | |
| Traction force, kN: main winch | 102 | 170 | 170 | 290 | 290 | |
| auxiliary winch | 37 | 80 | 80 | 140 | 140 | |
| Engine capacity, kW | 153 | 220 | 250 | 400 | 400 | |

| 1 | 2 | 3 | 4 | 5 | 6 |
|---|---|---|---|---|---|
| Extendable crawler track undercarriage: | | | | | |
| length, mm | 4660 | 5190 | 5260 | 6280 | 6560 |
| track width, mm | 700 (900) | 700 (900) | 750 | 900 | 900 |
| width, mm | 2500...4000 (2900...4200) | 2500...3900 (2700...4100) | 3100... 4450 | 3150.. .4700 | 3450... 5000 |
| Pressure on base, MPa | 0.070 (0.050) | 0.085 (0.080) | 0.098 | 0.070 | 0.120 |
| Overall dimensions in transportation position, mm: | | | | | |
| width | 2500 (2900) | 2500 (2700) | 3100 | 3150 | 3450 |
| height | 3190 | 3480 | 3630 | 3010 | 4170 |
| length | 16360 | 17370 | 18870 | 15180 | 18020 |
| Weight of rig, t: | | | | | |
| transportation | 33.5 | 48* | 50** | 53** | 107*; 79** |
| unladen | 39 | 55 | 75 | 105 | 140 |

*Notes:* 1. In the numerator, there are indicated the drilling parameters when an auger cleaner is used; in the denominator, there are indicated the drilling parameters when no cleaner is used. 2. Range of continuous flight augers is given in Table A.5.2.

\* Weight without counterweights.

\*\* Weight without counterweights and crawler-tracked undercarridge.

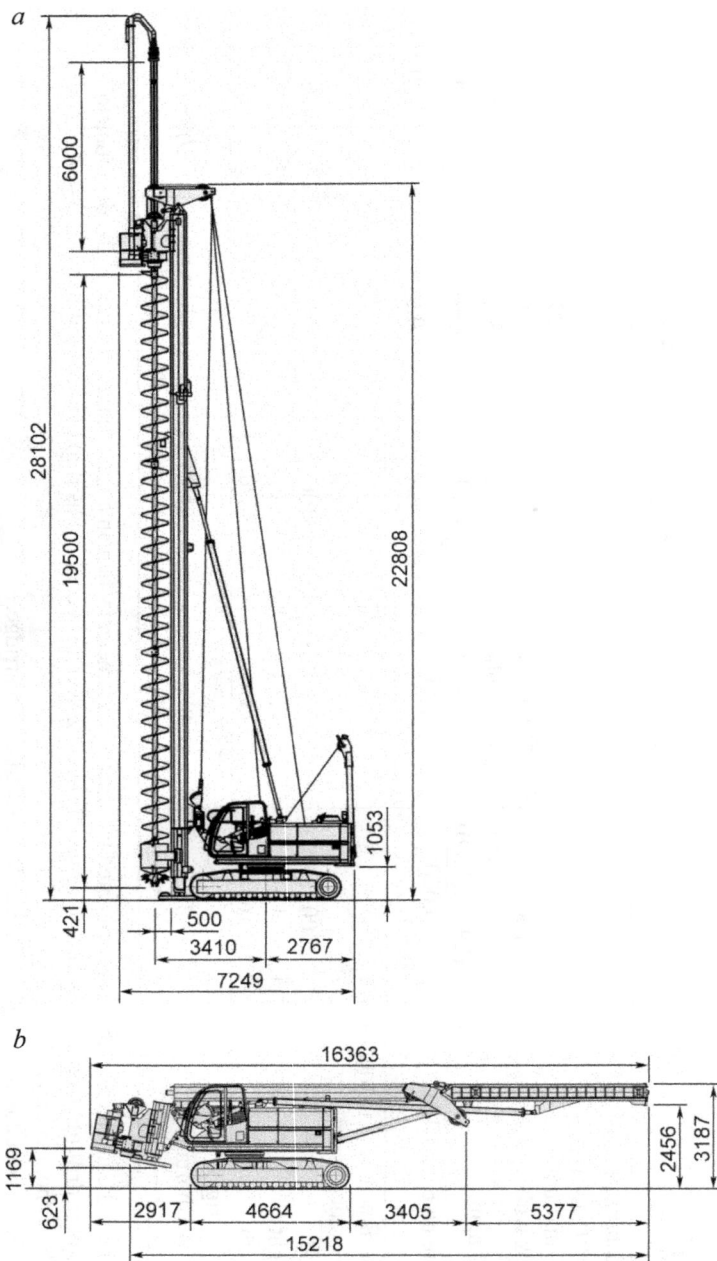

Figure A.5.1. SoilMec CM-50 rig to drill boreholes with continuous flight auger: operating (*a*) and transportation (*b*) position

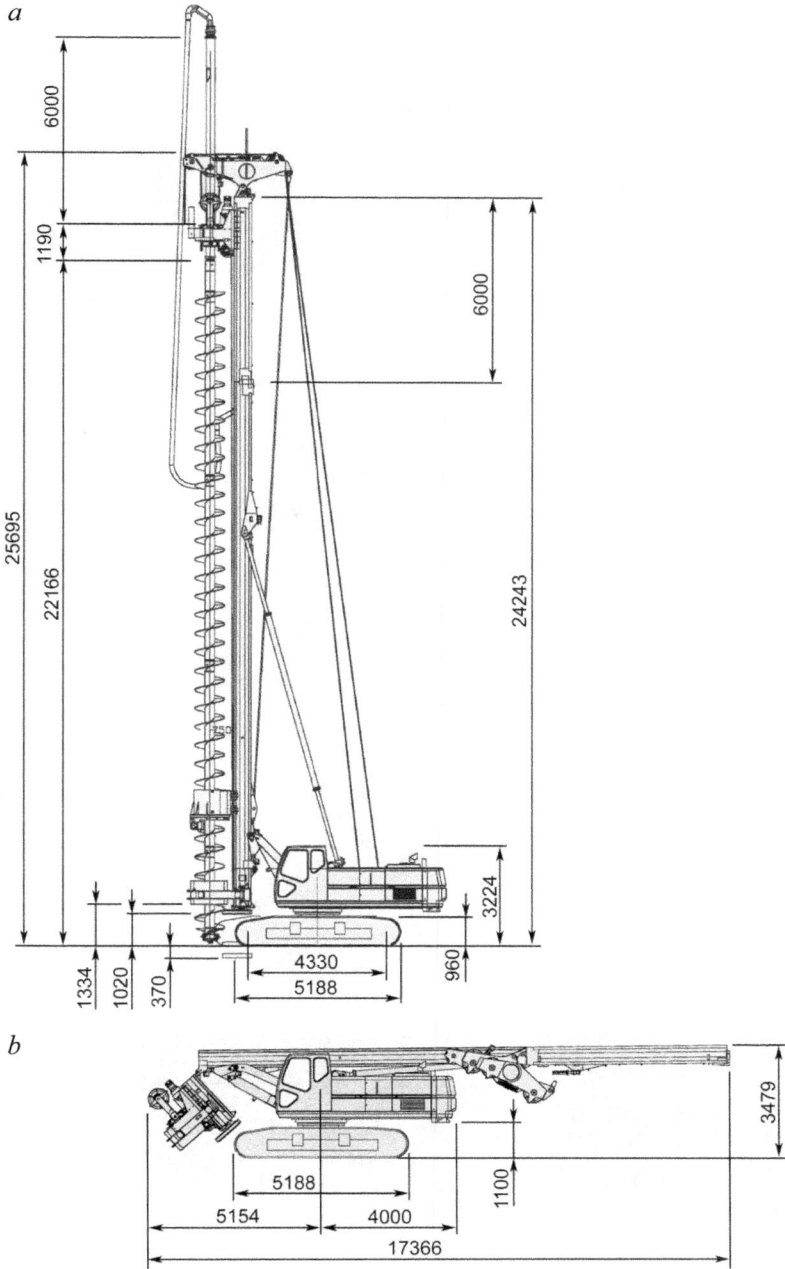

Figure A.5.2. SoilMec CM-70 rig to drill boreholes with continuous
flight auger: operating (*a*) and transportation (*b*) position

179

*a*

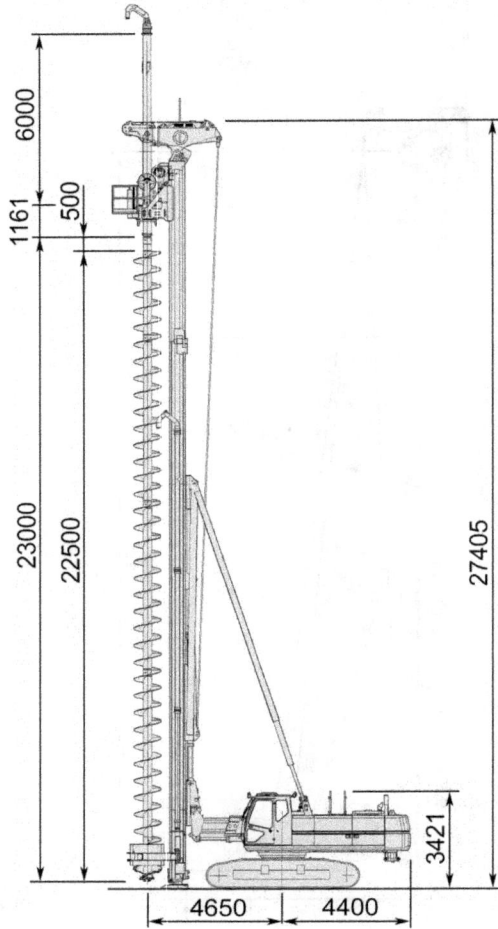

*b*

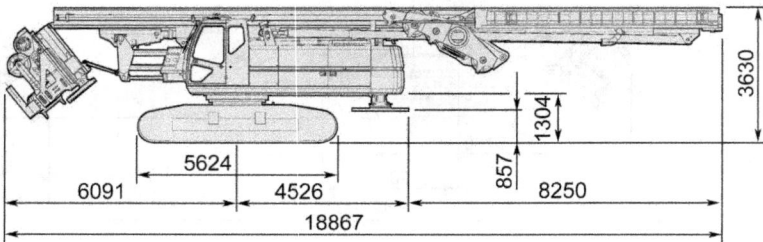

Figure A.5.3. SoilMec CM-700 rig to drill boreholes with continuous flight auger: operating (*a*) and transportation (*b*) position

*a*

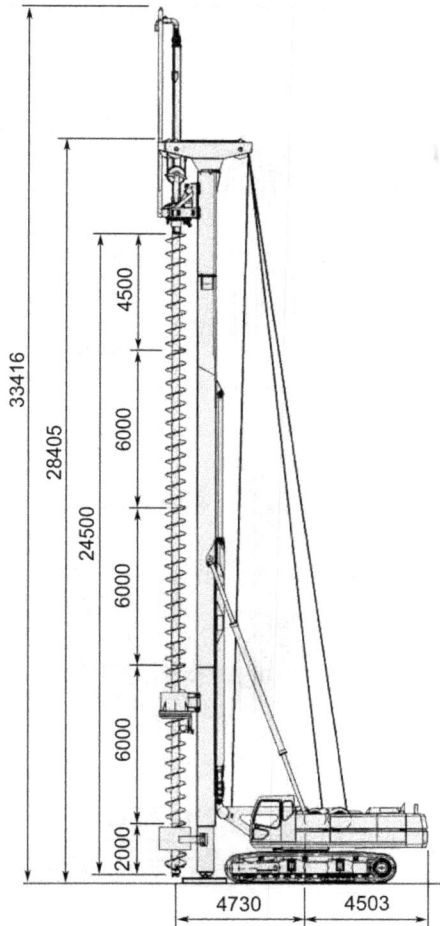

*b*

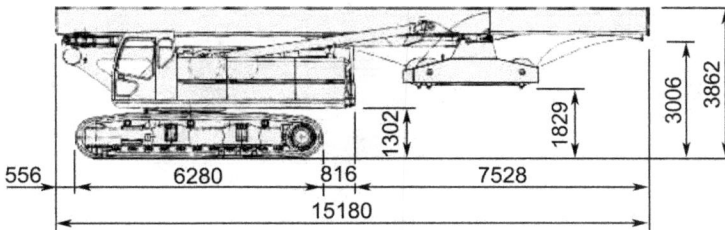

Figure A.5.4. SoilMec CM-120 rig to drill boreholes with continuous
flight auger: operating (*a*) and transportation (*b*) position

*a*

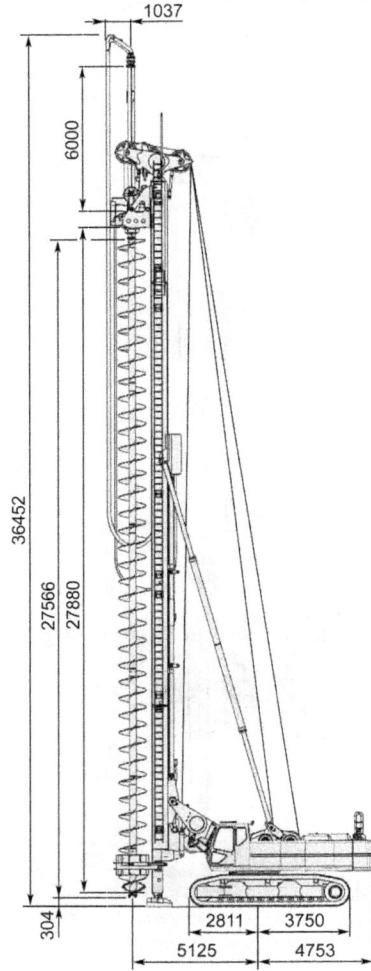

*b*

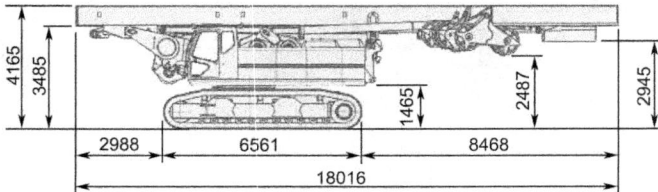

Figure A.5.5. SoilMec CM-1200 rig to drill boreholes with continuous flight auger: operating (*a*) and transportation (*b*) position

## Range of SoilMec continuous flight augers

| Auger diameter, mm | Type | | | | | | | |
| --- | --- | --- | --- | --- | --- | --- | --- | --- |
| | Auger with 4-inch internal passage diameter | | Auger with 5-inch internal passage diameter | | | | | |
| | HD-4 | | HD-5 | | XHD-5 | | 25HD-5 | |
| | Line | Pilot | Line | Pilot | Line | Pilot | Line | Pilot |
| 350 | + | + | | | | | | |
| 400 | + | + | | | | | | |
| 450 | + | + | + | + | | | | |
| 500 | + | + | + | + | | | | |
| 550 | + | + | + | + | | | | |
| 600 | + | + | + | + | + | + | + | + |
| 650 | + | + | + | + | + | + | + | + |
| 700 | + | + | + | + | + | + | + | + |
| 750 | + | + | + | + | + | + | + | + |
| 800 | + | + | + | + | + | + | + | + |
| 850 | | | | | + | + | + | + |
| 900 | | | | | + | + | + | + |
| 950 | | | | | + | + | + | + |
| 1000 | | | | | + | + | + | + |
| 1100 | | | | | | | + | + |
| 1200 | | | | | | | + | + |
| Auger length $L$, mm | 1500; 3000; 6000 | 1500 | 1500; 3000; 6000 | 1500 | 1500; 3000; 6000 | 1500 | 3000; 4500; 6000 | 2000 |

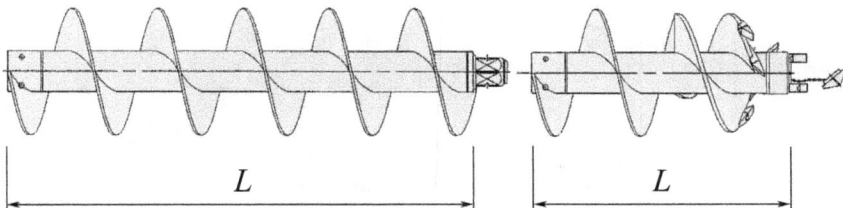

Figure A.5.6. Design of continuous flight auger

## Technical features of SoilMec R rigs to drill boreholes by Kelly method and with continuous flight auger

| Feature | Model | | | | | | | | | | |
|---|---|---|---|---|---|---|---|---|---|---|---|
| | R-210 | R-312/200 | R-416 | R-516 HD | R-620 | R-625 | R-725 | R-825 | R-930 | R-940 | R-1240 |
| 1 | 2 | 3 | 4 | 5 | 6 | 7 | 8 | 9 | 10 | 11 | 12 |
| Traction force, kN: | | | | | | | | | | | |
| main winch | 103 | 133 | 150 | 170 | 192 | 240 | 240 | 240 | 290 | 320 | 320 |
| auxiliary winch | 41 | 56 | 65 | 75 | 80 | 140 | 140 | 140 | 140 | 140 | 140 |
| Maximum torsion torque, kN·m | 100 | 130 | 160 | 180 | 200 | 240 | 240 | 240 | 305 | 469 | 469 |
| Maximum rotation speed of drilling tool, rpm | | | | | | | | | | | |
| when drilling | 43 | 42 | 30 | 30 | 34 | 30 | 38 | 28 | 25 | 19 | 19 |
| when pulling out | 147 | 153 | 160 | 137 | 144 | 140 | 136 | 140 | 117 | 53 | 53 |
| Engine capacity, kW | 116 | 150 | 230 | 230 | 260 | 300 | 330 | 300 | 400 | 400 | 400 |
| Extendable crawler track undercarriage: | | | | | | | | | | | |
| length, mm | 3800 | 4500 | 5200 | 5200 | 5300 | 5700 | 5700 | 6000 | 6000 | 6100 | 6100 |
| track width, mm | 600 | 600 | 700 | 700 | 750 | 900 | 900 | 900 | 900 | 900 | 900 |
| minimum width, mm | 2550 | 2540 | 2900 | 2500 | 3000 | 2980 | 3180 | 3400 | 3400 | 3400 | 3450 |
| maximum width, mm | 3400 | 3700 | 4300 | 3900 | 4500 | 4500 | 4700 | 5000 | 5000 | 5000 | 5000 |
| Transportation weight of rig, t | 26 | 34 | 42 | 36...39 | 58 | 58 | 60 | 36 | 45 | 48 | 65 |

| 1 | 2 | 3 | 4 | 5 | 6 | 7 | 8 | 9 | 10 | 11 | 12 |
|---|---|---|---|---|---|---|---|---|----|----|----|
| *Borehole drilling parameters with continuous auger* | | | | | | | | | | | |
| Maximum drilling diameter, mm | 750 | 750 | 1000 | 1000 | 1000 | 1200 | 1200 | 1200 | 1200 | – | 1200 |
| Maximum drilling depth without auger extension, m | 10.5 | 13 | 14.5 | 15 | 16.5 | 19.5 | 16 | 21 | 22.6 | – | 17.5 |
| Maximum drilling depth with auger extension, m | 15.3 | 19 | 20.5 | 18 | 22.5 | 25.5 | 24 | 27 | 28.6 | – | 23.5 |
| Auger pulling force, kN | 140, 280 | 380 | 500 | 520 | 600 | 732 | 800 | 732 | 1160 | – | 1060 |
| Unladen weight of rig, t | 27 | 34 | 48 | 50...53 | 60...63 | 70 | 80 | 80...85 | 95...105 | – | 140 |
| Overall dimensions in transportation position, mm: | | | | | | | | | | | |
| width | 2550 | 2540 | 2900 | 2500 | 3000 | 2980 | 3180 | 3400 | 3400 | 3400 | 3450 |
| height | – | 3426 | 3197 | 3300 | 3500 | 3518 | – | 3566 | – | – | – |
| length | – | 12119 | 13064 | 14234 | 14181 | 17514 | – | 17563 | – | – | – |
| *Borehole drilling parameters by Kelly method* | | | | | | | | | | | |
| Maximum drilling diameter, mm: | | | | | | | | | | | |
| without casing pipe | 1200 | 1500 | 1500 | 2000 | 1800 | 2500 | 2000 | 2500 | 3000 | 3000 | 2750 |
| with casing pipe | | | 1300 | 1500 | 1500 | 2000 | 2000 | 2000 | 2500 | 2500 | 2500 |
| Maximum drilling depth, m | 40 | 48 | 56 | 61 | 66 | 77 | 77.4 | 77.4 | 76.5 | 92 | 86.5 |

| 1 | 2 | 3 | 4 | 5 | 6 | 7 | 8 | 9 | 10 | 11 | 12 |
|---|---|---|---|---|---|---|---|---|---|---|---|
| Cylinder stroke, m | 3.0 | 3.5 | 5.1 | 5.1 | 5.6 | 6.5 | — | 6.5 | 6.5 | 6.5 | — |
| Feed force of drilling tool when using cylinder, kN: | | | | | | | | | | | |
| upward | 124 | 124 | 154 | 185 | 210 | 264 | — | 264 | 320 | 416 | — |
| downward | 68 | 102 | 120 | 122 | 150 | 201 | — | 201 | 153 | 330 | — |
| Downward stroke of drilling tool at winch-using feed, m | — | — | 11.0 | — | 13.0 | 12.5 | 16.3 | 12.5 | 13.5 | 14.0 | 17.5 |
| Downfeed force of drilling tool when using winch, kN | — | — | 165 | — | 200 | 200 | 320 | 200 | 360 | 360 | 420 |
| Unladen weight of rig, t | 27 | 35 | 49 | 50...55 | 63 | 70...72 | 80 | 80...85 | 105...110 | 115...120 | 140 |
| Pressure on soil, MPa | 0.073 | 0.078 | 0.080 | 0.083 | 0.090 | 0.080 | 0.095 | 0.090 | 0.100 | 0.130 | 0.150 |
| Overall dimensions in transportation position, mm: | | | | | | | | | | | |
| width | 2550 | 2540 | 2900 | 2500 | 3000 | 2980 | 3180 | 3400 | 3400 | 3400 | 3450 |
| height | 3053 | 3277 | 3197 | 3300 | 3324 | — | 3717 | 3524 | 3910 | 3440 | 3342 |
| length | 12431 | 12090 | 12758 | 13794 | 14181 | — | 15494 | 13710 | 15756 | 16255 | 16954 |

*Notes:* 1. SoilMec produces short augers to drill boreholes 400 to 2000 mm in diameter (at an interval of 100 mm), rotary core bits and bucket augers to drill boreholes 600 to 2000 mm to drill boreholes 310 to 610 mm (at an interval of 100 mm), helical tools to ream boreholes 310 to 610 mm (at an interval of 50 mm). 2. Range of continuous flight augers is given in Table A.5.2. 3. Depth of borehole drilling by Kelly method (depending on telescopic bars used) is given in Table A.5.4.

*a*

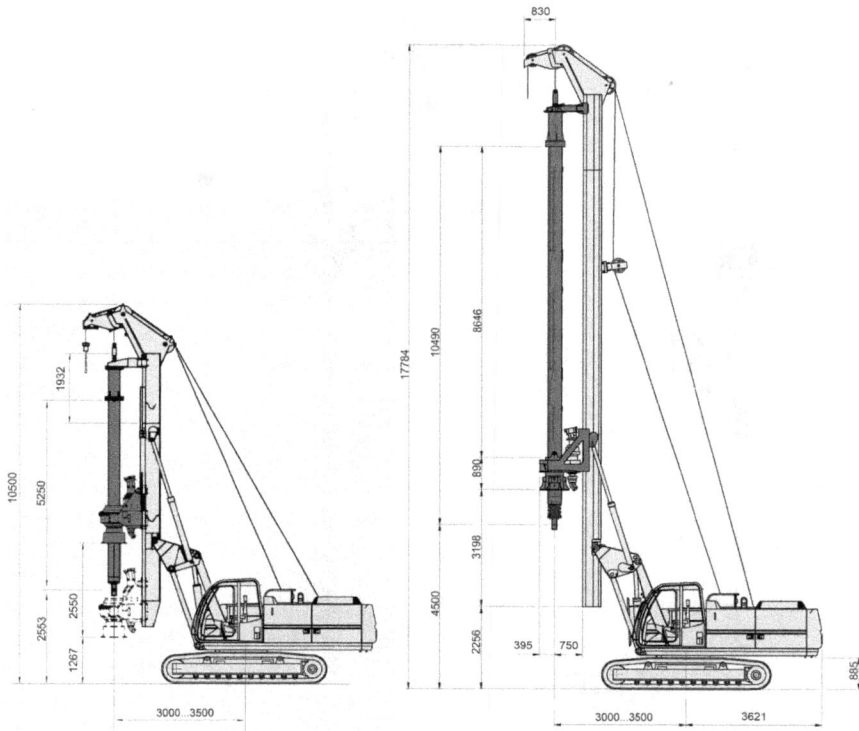

*b*

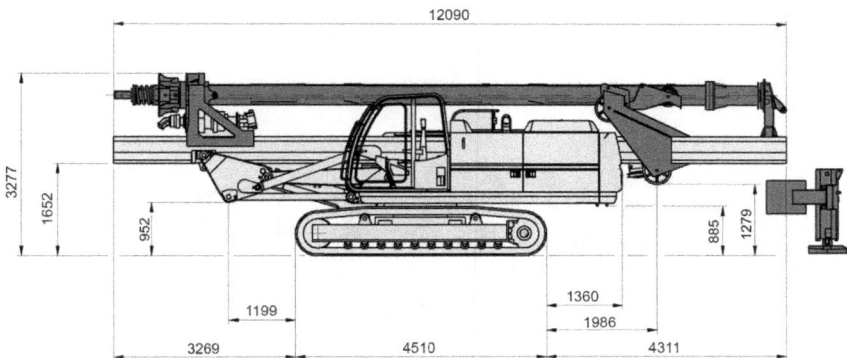

Figure A.5.7. SoilMec R-312/200 rig with equipment to drill boreholes by Kelly method: operating (*a*) and transportation (*b*) position

*a*

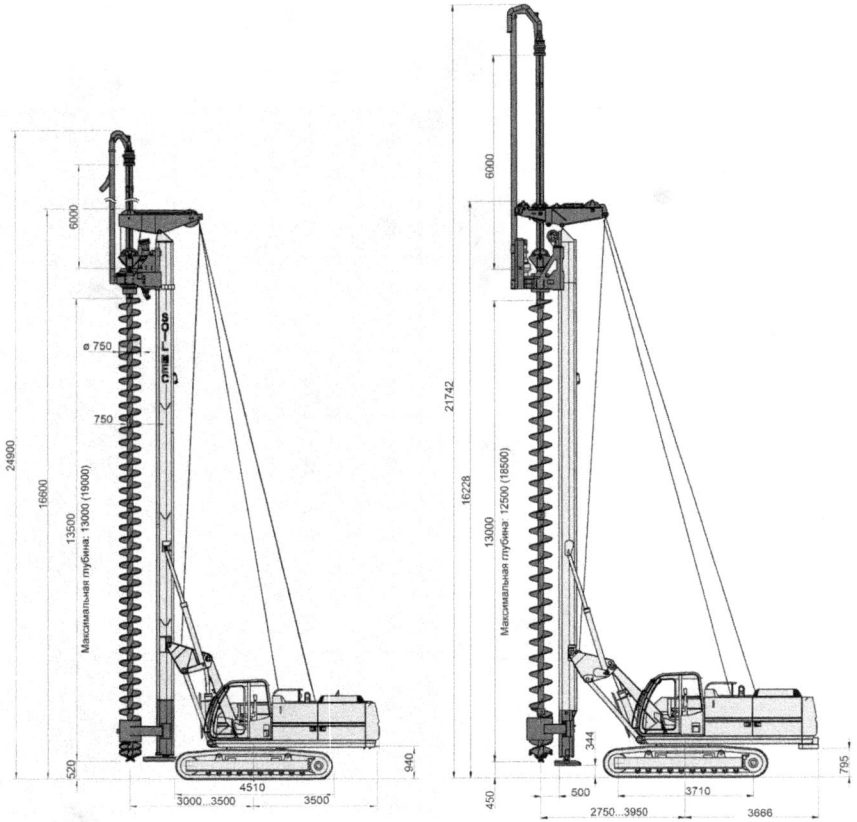

*b*

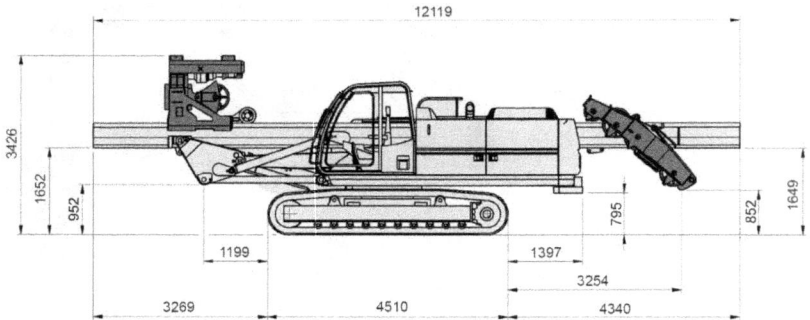

Figure A.5.8. SoilMec R-312/200 rig with equipment designed for drilling bore-holes with continuous flight auger: operating (*a*) and transportation (*b*) positions

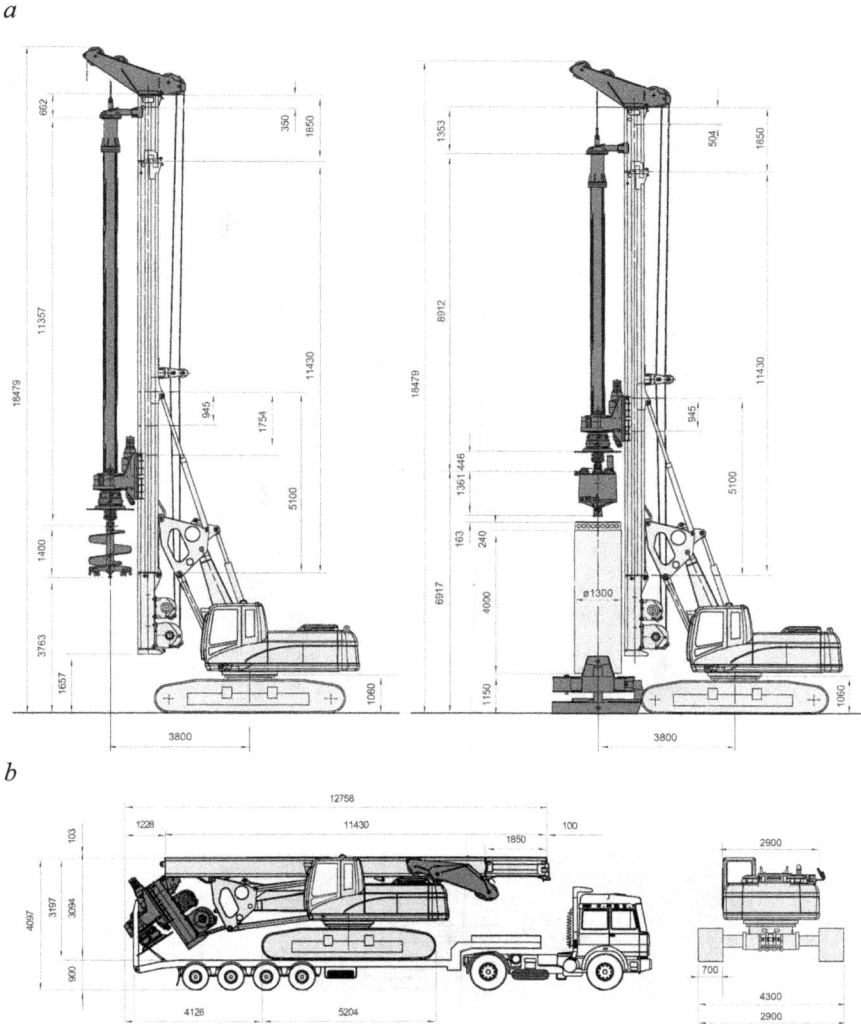

Figure A.5.9. SoilMec R-416 rig with equipment to drill boreholes by Kelly method: operating (*a*) and transportation (*b*) positions

*a*

*b*

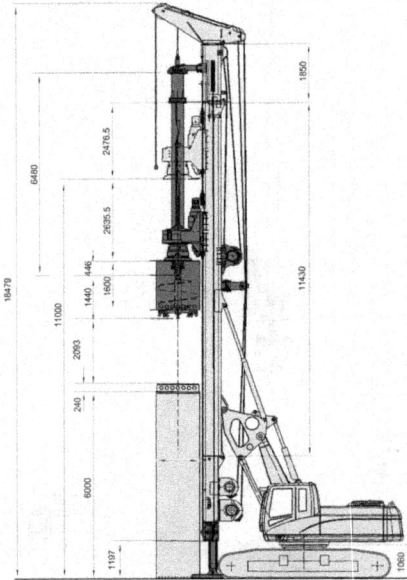
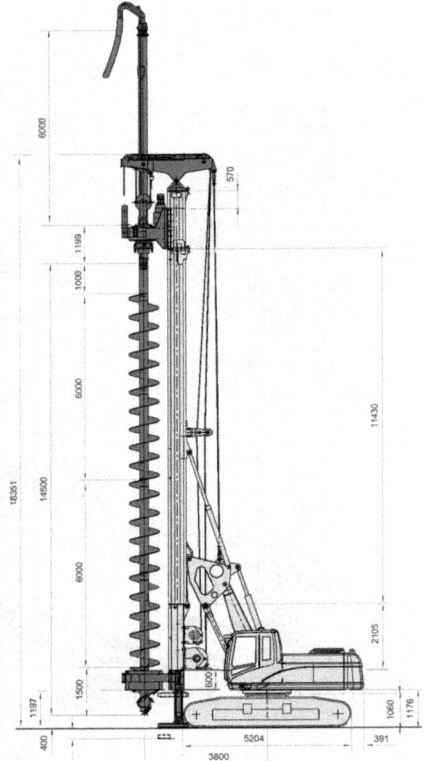

*c*

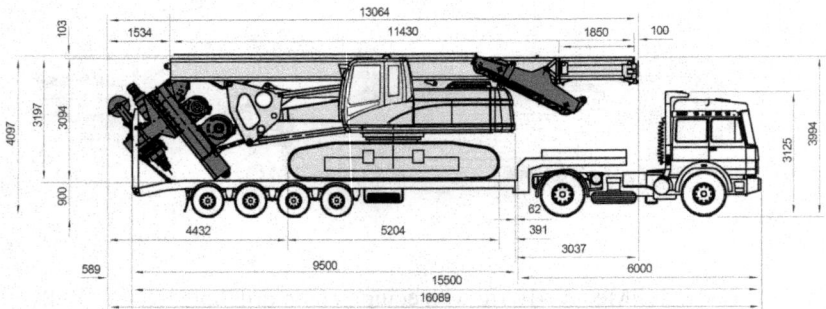

Figure A.5.10. SoilMec R-416 rig: operating position of rig with equipment to drill boreholes by Kelly method (*a*); operating (*b*) and transportation (*c*) position of rig with equipment designed for drilling boreholes with continuous flight auger

*a*

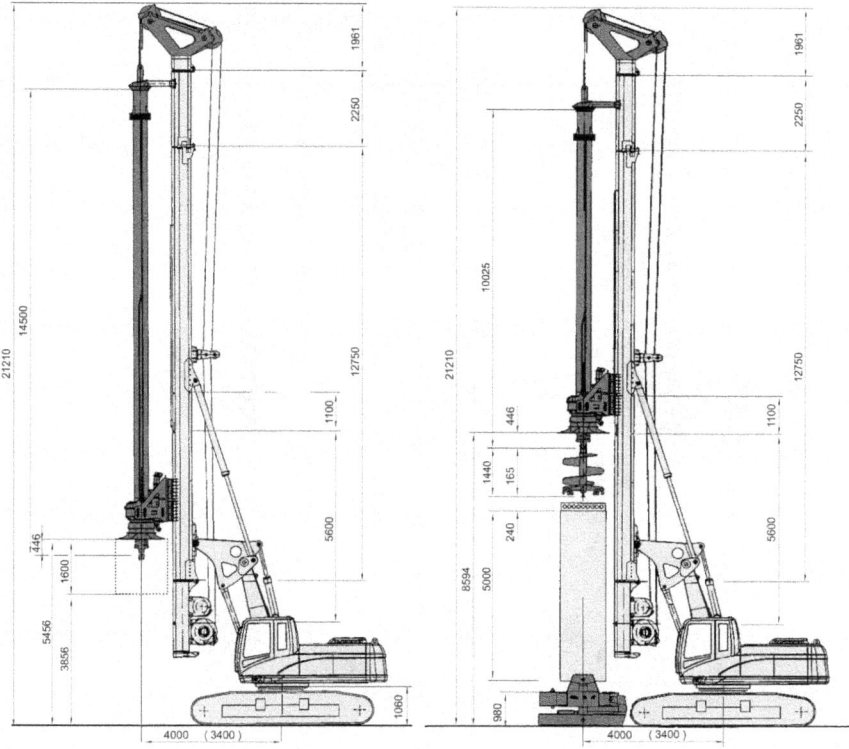

*b*

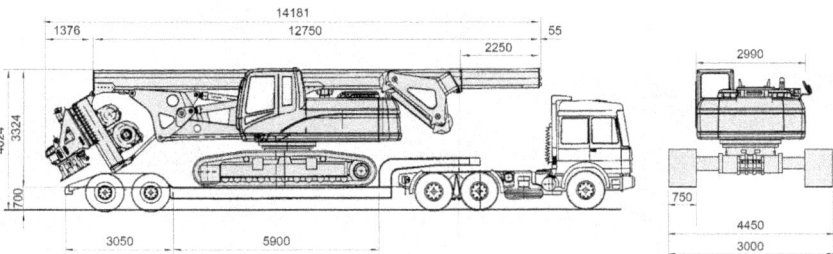

Figure A.5.11. SoilMec R-620 rig with equipment to drill boreholes by Kelly method: operating (*a*) and transportation (*b*) positions

*a*

*b*

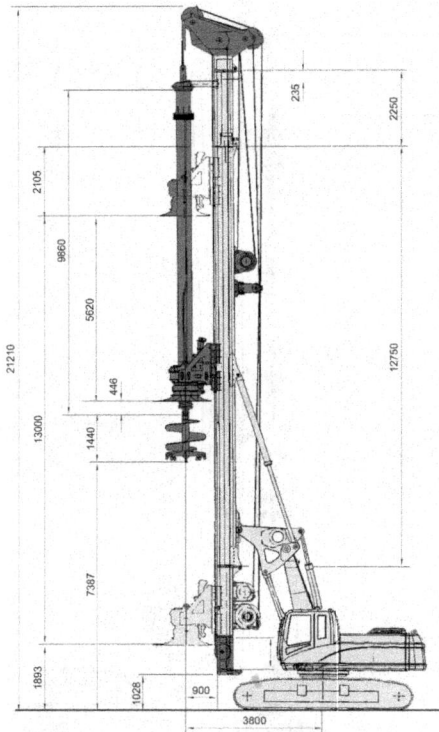

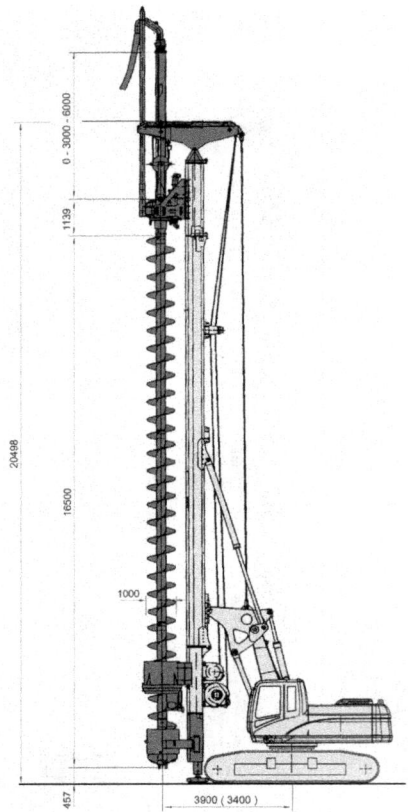

*c*

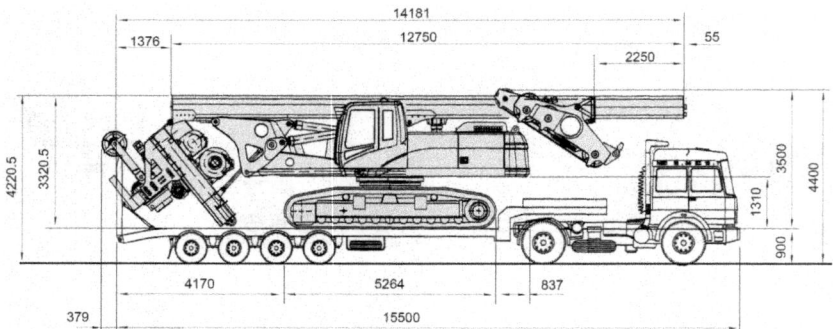

Figure A.5.12. SoilMec R-620 rig: operating position of rig with equipment to drill boreholes by Kelly method (*a*); operating (*b*) and transportation (*c*) position of rig with equipment designed for drilling boreholes with continuous flight auger

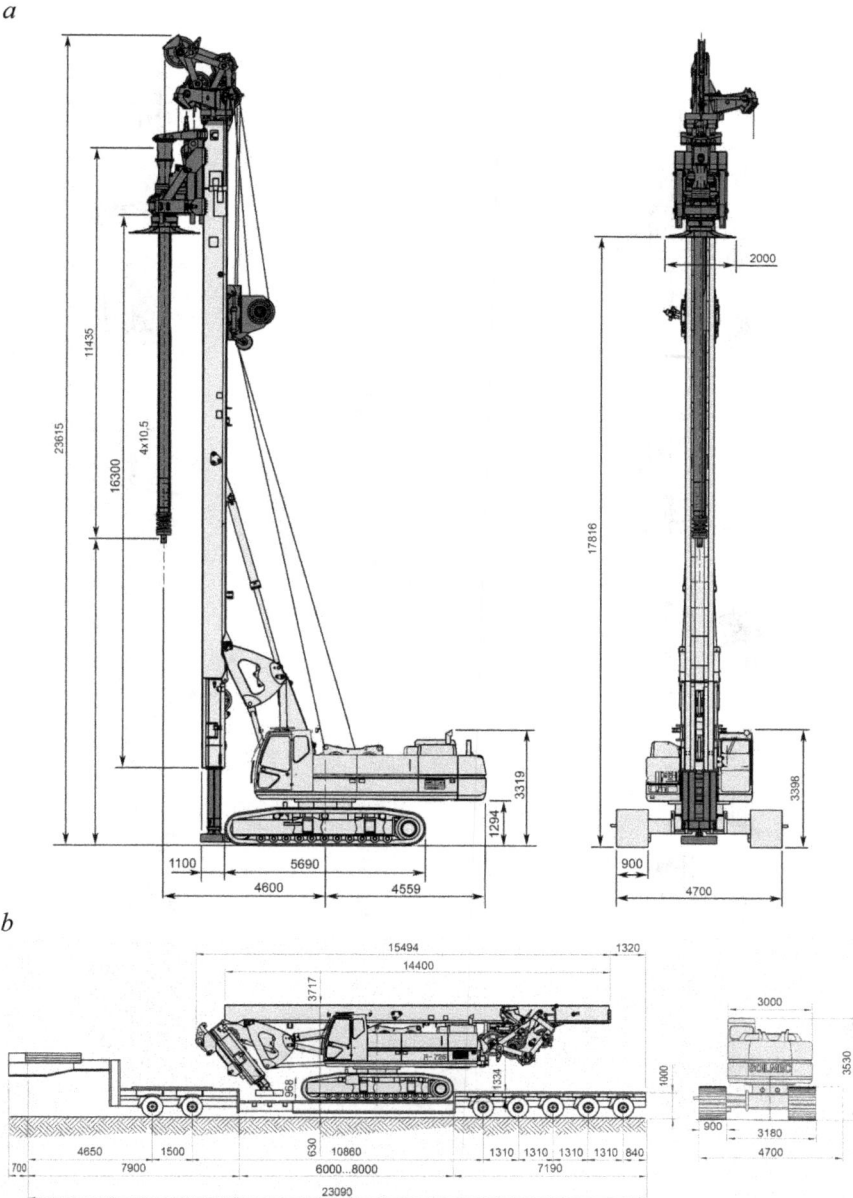

Figure A.5.13. SoilMec R-725 rig with equipment to drill boreholes by Kelly method: operating (*a*) and transportation (*b*) positions

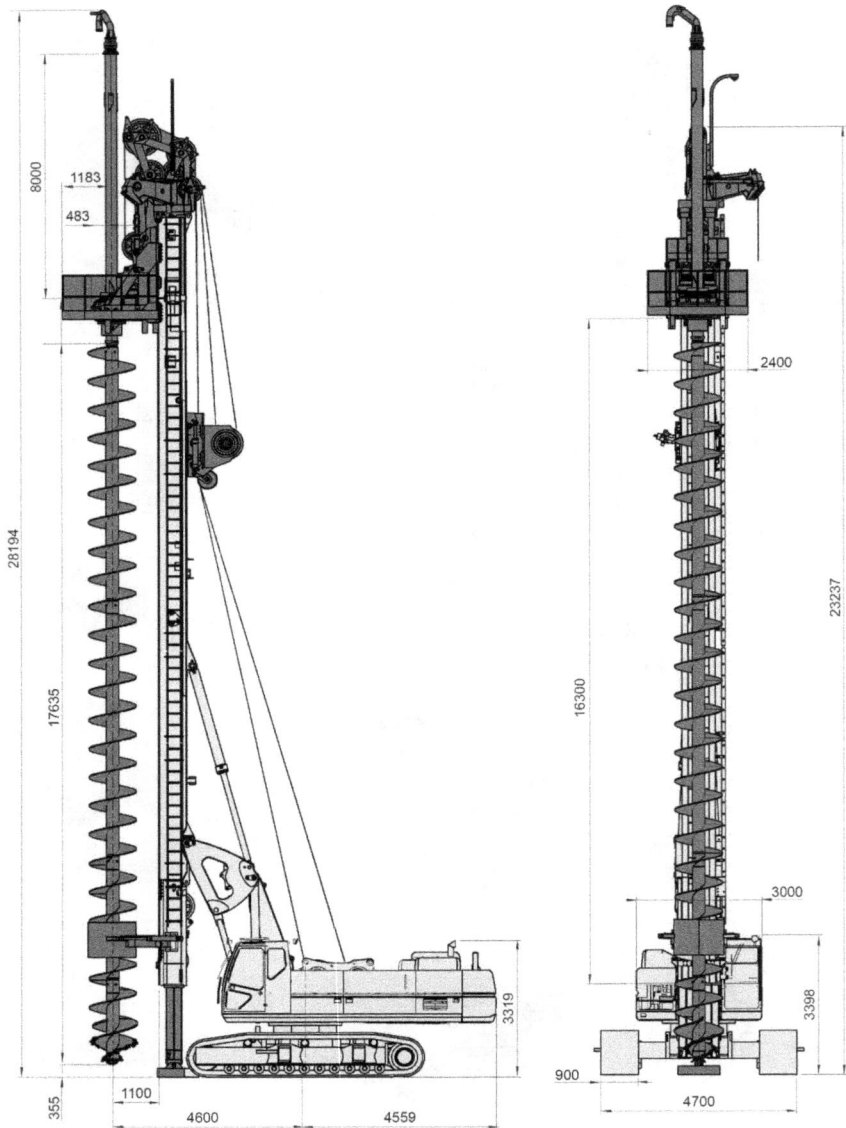

Figure A.5.14. SoilMec R-725 rig with equipment
to drill boreholes with continuous flight auger

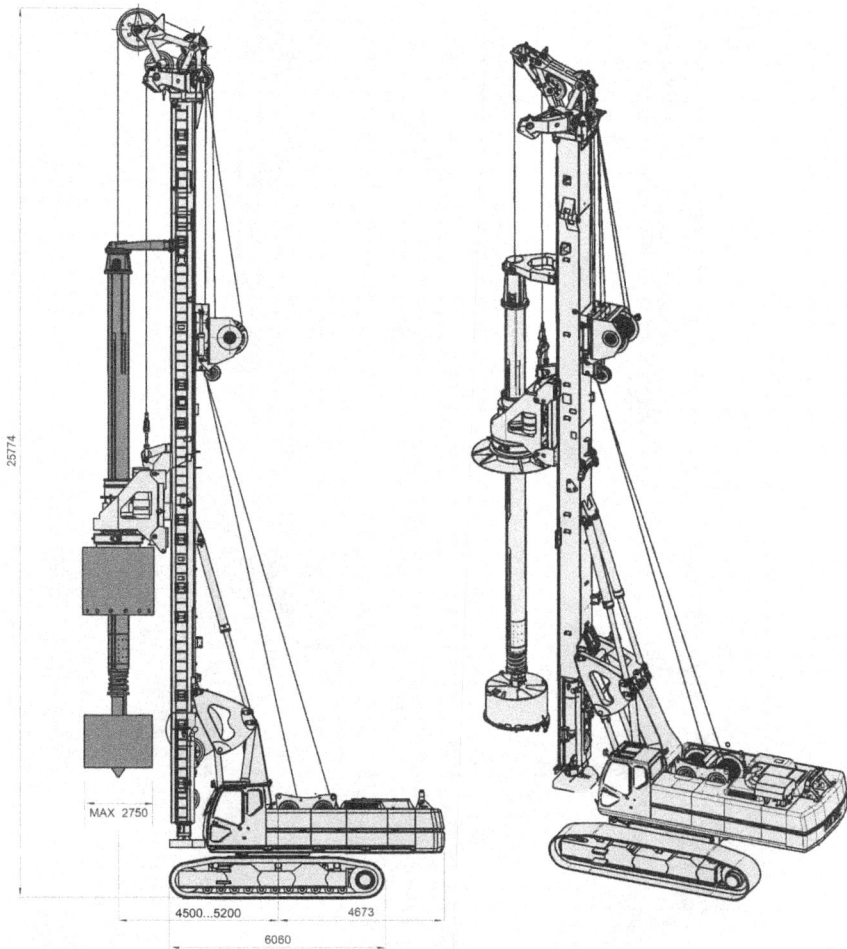

Figure A.5.15. SoilMec R-1240 rig with equipment
to drill boreholes by Kelly method

*a*

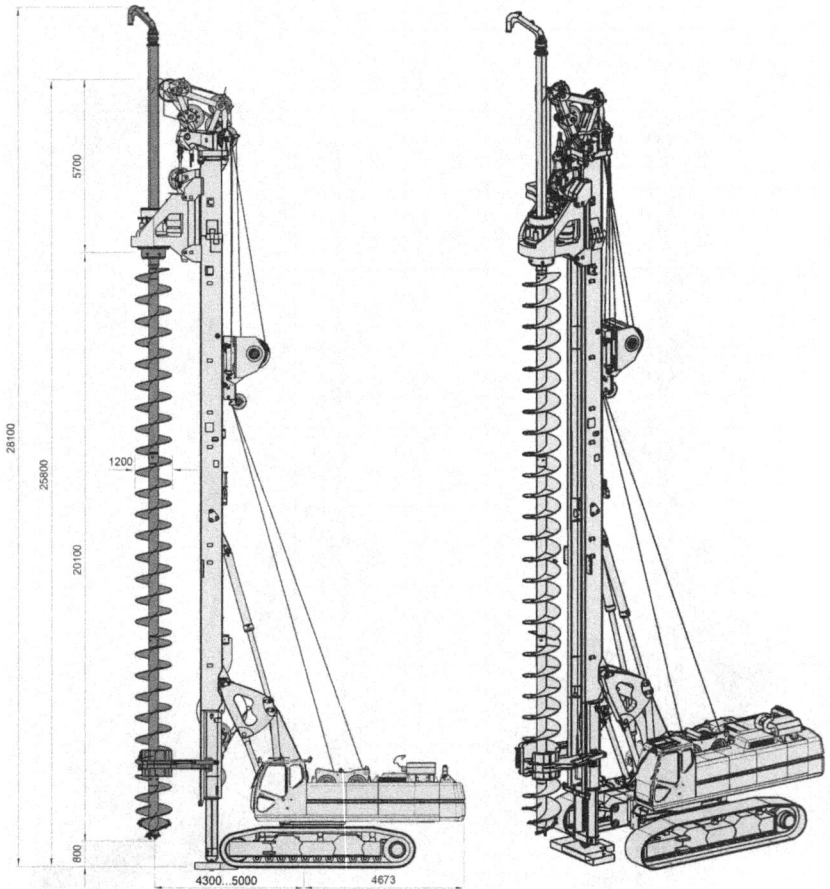

*b*

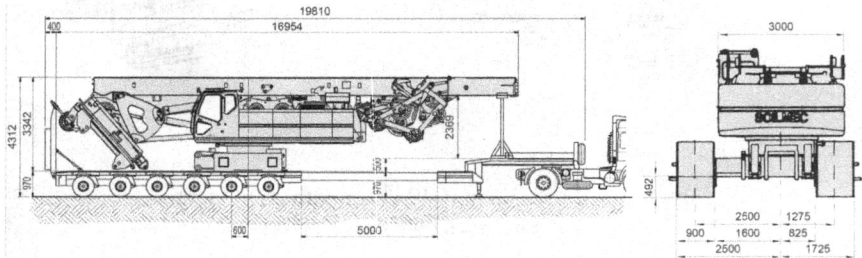

Figure A.5.16. SoilMec R-1240 rig with equipment designed for drilling boreholes with continuous flight auger: operating (*a*) and transportation (*b*) positions

## Depth of borehole drilling by Kelly method

| Model of SoilMec rig | Features of SoilMec telescopic bar to drill boreholes by Kelly method | | | | | | |
|---|---|---|---|---|---|---|---|
| | Type | $D$, mm | $n$, ea. | $L$, m | $h$, m | $m$, t | $H$, m |
| 1 | 2 | 3 | 4 | 5 | 6 | 7 | 8 |
| R-312/200 | Locking-type bar | 355 | 4 | 7.5 | 5.0 | 2.9 | 25.0 |
| | | | 4 | 9.0 | 5.0 | 3.3 | 34.0 |
| | | | 4 | 11.0 | 4.6 | 3.9 | 38.0 |
| | Friction-type bar | 355 | 4 | 6.0 | 5.0 | 2.4 | 19.0 |
| | | | 4 | 7.5 | 5.0 | 2.9 | 25.0 |
| | | | 4 | 9.0 | 5.0 | 3.3 | 32.0 |
| | | | 4 | 11.0 | 4.6 | 3.9 | 39.0 |
| | | | 5 | 9.0 | 5.0 | 4.3 | 40.0 |
| | | | 5 | 11.0 | 4.6 | 5.1 | 48.0 |
| R-416 | Locking-type bar | 355 | 4 | 6.0 | 6.9 | 2.4 | 18.5 |
| | | | 4 | 8.0 | 6.9 | 3.0 | 28.5 |
| | | | 4 | 9.0 | 6.6 | 3.3 | 32.5 |
| | | | 4 | 10.5 | 5.1 | 3.8 | 38.5 |
| | | | 4 | 12.0 | 3.4 | 4.2 | 45.0 |
| | | | 4 | 13.0 | 2.6 | 4.6 | 48.0 |
| | Friction-type bar | 355 | 4 | 6.0 | 6.9 | 2.4 | 20.0 |
| | | | 4 | 9.0 | 6.6 | 3.4 | 33.0 |
| | | | 4 | 11.0 | 4.2 | 3.7 | 40.0 |
| | | | 4 | 12.0 | 3.4 | 4.2 | 46.0 |
| | | | 5 | 8.5 | 6.9 | 4.1 | 39.0 |
| | | | 5 | 9.0 | 6.6 | 4.3 | 41.0 |
| | | | 5 | 11.0 | 3.6 | 5.1 | 50.0 |
| | | | 5 | 12.0 | 3.8 | 5.5 | 56.0 |
| R-516HD | Locking-type bar | 406 | 4 | 8.5 | 8.1 | 5.1 | 29.5 |
| | | | 4 | 9.0 | 7.7 | 5.3 | 31.8 |
| | | | 4 | 11.0 | 5.8 | 6.3 | 38.7 |
| | | | 4 | 12.5 | 4.3 | 7.0 | 45.3 |
| | Friction-type bar | 406 | 5 | 9.0 | 7.6 | 5.5 | 39.5 |
| | | | 5 | 12.0 | 4.5 | 7.0 | 55.0 |
| | | | 5 | 13.0 | 3.4 | 7.5 | 61.0 |
| R-620 | Locking-type bar | 406 | 4 | 9.0 | 8.1 | 5.2 | 32.0 |
| | | | 4 | 10.0 | 7.9 | 5.6 | 35.0 |
| | | | 4 | 12.0 | 6.2 | 6.4 | 42.0 |
| | | | 4 | 13.5 | 4.4 | 7.2 | 49.0 |
| | | | 4 | 14.5 | 3.5 | 7.7 | 53.0 |
| | Friction-type bar | 406 | 4 | 9.0 | 8.1 | 4.0 | 32.0 |
| | | | 4 | 10.5 | 7.6 | 5.7 | 37.0 |
| | | | 4 | 11.5 | 6.4 | 6.3 | 41.5 |
| | | | 4 | 13.5 | 4.4 | 7.1 | 50.0 |
| | | | 5 | 9.0 | 8.1 | 4.9 | 39.5 |
| | | | 5 | 10.5 | 7.6 | 5.5 | 46.0 |
| | | | 5 | 12.0 | 5.9 | 6.4 | 55.0 |
| | | | 5 | 13.5 | 4.5 | 6.9 | 62.0 |
| | | | 5 | 14.0 | 4.2 | 7.1 | 63.5 |
| | | | 5 | 14.5 | 3.6 | 7.4 | 66.0 |

| 1 | 2 | 3 | 4 | 5 | 6 | 7 | 8 |
|---|---|---|---|---|---|---|---|
| R-725 | Locking-type bar | 406 | 4 | 10.5 | 8.9 | 5.8 | 37.0 |
| | | | 4 | 11.5 | 7.8 | 6.3 | 41.1 |
| | | | 4 | 13.5 | 5.7 | 7.2 | 49.6 |
| | | | 4 | 15.5 | 3.7 | 8.2 | 57.8 |
| | | | 4 | 16.5 | 2.6 | 8.7 | 62.0 |
| | Friction-type bar | 406 | 4 | 10.5 | 8.9 | 5.7 | 37.0 |
| | | | 4 | 11.5 | 7.8 | 6.2 | 41.5 |
| | | | 4 | 13.5 | 5.7 | 7.1 | 49.4 |
| | | | 4 | 15.5 | 3.7 | 8.0 | 57.9 |
| | | | 4 | 16.5 | 2.6 | 8.5 | 62.0 |
| | | | 5 | 10.5 | 8.9 | 5.5 | 46.2 |
| | | | 5 | 11.5 | 7.8 | 6.0 | 51.6 |
| | | | 5 | 13.5 | 5.7 | 6.9 | 61.9 |
| | | | 5 | 15.5 | 3.7 | 7.9 | 72.2 |
| | | | 5 | 16.5 | 2.6 | 6.4 | 77.4 |
| R-825 | Locking-type bar | 406 | 4 | 10.5 | 9.2 | 5.7 | 37.0 |
| | | | 4 | 11.5 | 8.1 | 6.2 | 41.5 |
| | | | 4 | 13.5 | 6.0 | 7.1 | 49.4 |
| | | | 4 | 15.5 | 3.9 | 8.0 | 57.9 |
| | | | 4 | 16.5 | 2.9 | 8.5 | 62.0 |
| | | | 5 | 10.5 | 9.2 | 5.5 | 46.2 |
| | | | 5 | 11.5 | 8.1 | 6.0 | 51.6 |
| | | | 5 | 13.5 | 6.0 | 6.9 | 61.9 |
| | | | 5 | 15.5 | 3.9 | 7.9 | 72.2 |
| | | | 5 | 16.5 | 2.9 | 8.4 | 77.4 |
| | Friction-type bar | 406 | 4 | 10.5 | 9.2 | 5.8 | 37.0 |
| | | | 4 | 11.5 | 8.1 | 6.3 | 41.1 |
| | | | 4 | 13.5 | 6.0 | 7.2 | 49.6 |
| | | | 4 | 15.5 | 3.9 | 8.2 | 57.8 |
| | | | 4 | 16.5 | 2.9 | 8.7 | 62.0 |
| R-930 | Locking-type bar | 483 | 4 | 10.5 | 8.9 | 7.0 | 37.0 |
| | | | 4 | 11.5 | 8.8 | 7.5 | 41.0 |
| | | | 4 | 13.5 | 6.8 | 8.5 | 49.0 |
| | | | 4 | 15.5 | 4.8 | 9.5 | 57.0 |
| | | | 4 | 16.5 | 3.9 | 10.0 | 61.0 |
| | Friction-type bar | 483 | 4 | 10.5 | 8.9 | 7.0 | 37.0 |
| | | | 4 | 11.5 | 8.8 | 7.5 | 41.0 |
| | | | 4 | 13.5 | 6.8 | 8.5 | 49.5 |
| | | | 4 | 15.5 | 4.8 | 9.5 | 57.5 |
| | | | 4 | 16.5 | 3.9 | 10.0 | 61.0 |
| | | | 5 | 10.5 | 9.9 | 7.6 | 46.0 |
| | | | 5 | 11.5 | 8.8 | 8.3 | 51.5 |
| | | | 5 | 13.5 | 6.8 | 9.6 | 62.0 |
| | | | 5 | 15.5 | 4.8 | 10.9 | 72.0 |
| | | | 5 | 16.5 | 3.9 | 11.5 | 76.5 |

| 1 | 2 | 3 | 4 | 5 | 6 | 7 | 8 |
|---|---|---|---|---|---|---|---|
| R-940 | Locking-type bar | 559 | 4 | 10.5 | 8.3 | 8.5 | 37.0 |
| | | | 4 | 11.5 | 8.3 | 9.0 | 41.0 |
| | | | 4 | 13.5 | 6.8 | 10.3 | 49.0 |
| | | | 4 | 15.5 | 4.8 | 11.5 | 57.0 |
| | | | 4 | 16.5 | 3.9 | 12.2 | 61.0 |
| | | | 5 | 10.5 | 8.3 | 8.5 | 46.5 |
| | | | 5 | 11.5 | 8.3 | 9.0 | 51.5 |
| | | | 5 | 13.5 | 6.8 | 10.3 | 62.0 |
| | | | 5 | 15.5 | 4.8 | 11.5 | 72.0 |
| | | | 5 | 16.5 | 3.9 | 12.2 | 76.5 |
| | Friction-type bar | 559 | 4 | 10.5 | 8.3 | 8.5 | 37.0 |
| | | | 4 | 11.5 | 8.3 | 9.0 | 41.0 |
| | | | 4 | 13.5 | 6.8 | 10.3 | 49.5 |
| | | | 4 | 15.5 | 4.8 | 11.5 | 57.5 |
| | | | 4 | 16.5 | 3.9 | 12.2 | 61.0 |
| | | | 5 | 10.5 | 8.3 | 10.0 | 46.0 |
| | | | 5 | 11.5 | 8.3 | 10.8 | 51.5 |
| | | | 5 | 13.5 | 6.8 | 12.4 | 62.0 |
| | | | 5 | 15.5 | 4.8 | 14.0 | 72.0 |
| | | | 5 | 16.5 | 3.9 | 14.7 | 77.0 |
| | | | 6 | 10.5 | 8.3 | 10.3 | 56.0 |
| | | | 6 | 11.5 | 8.3 | 11.1 | 62.0 |
| | | | 6 | 13.5 | 6.8 | 12.8 | 74.5 |
| | | | 6 | 15.5 | 4.8 | 14.5 | 86.5 |
| | | | 6 | 16.5 | 3.9 | 15.4 | 92.0 |
| R-1240 | Locking-type bar | 559 | 4 | 10.5 | 10.8 | 8.5 | 37.0 |
| | | | 4 | 11.5 | 9.7 | 9.0 | 41.0 |
| | | | 4 | 13.5 | 7.7 | 10.3 | 49.5 |
| | | | 4 | 15.5 | 5.7 | 11.5 | 57.5 |
| | | | 5 | 10.5 | 10.8 | 10.0 | 46.5 |
| | | | 5 | 11.5 | 9.7 | 10.8 | 51.5 |
| | | | 5 | 13.5 | 7.7 | 12.4 | 62.0 |
| | | | 5 | 15.5 | 5.7 | 14.0 | 72.0 |
| | Friction-type bar | 559 | 4 | 10.5 | 10.8 | 8.5 | 37.0 |
| | | | 4 | 11.5 | 9.7 | 9.0 | 41.0 |
| | | | 4 | 13.5 | 7.7 | 10.3 | 49.5 |
| | | | 4 | 15.5 | 5.7 | 11.5 | 57.5 |
| | | | 5 | 10.5 | 10.8 | 10.0 | 46.5 |
| | | | 5 | 11.5 | 9.7 | 10.8 | 51.5 |
| | | | 5 | 13.5 | 7.7 | 12.4 | 62.0 |
| | | | 5 | 15.5 | 5.7 | 14.0 | 72.0 |
| | | | 6 | 10.5 | 10.8 | 10.3 | 56.0 |
| | | | 6 | 11.5 | 9.7 | 11.1 | 62.0 |
| | | | 6 | 13.5 | 7.7 | 12.8 | 74.5 |
| | | | 6 | 15.5 | 5.7 | 14.5 | 86.5 |

*Note.* The following abbreviations are taken in the table: $D$ = outside diameter of bar; $n$ = number of Kelly bar sections; $L$ = length of bar sections; $m$ = weight of Kelly bar; $h$ = location height of rotary swivel (mounting attachment of Kelly bar) above ground level; $H$ = depth of drilling by Kelly method.

*Table A.5.5*

## Technical features of SoilMec concrete pumps

| Feature | Model | | | | |
|---|---|---|---|---|---|
| | P4.65 | P6.80 | P6.90 | P6.100 | P6.120 |
| Nominal power, kW | 88 | 110 | 132 | 132 | 132 |
| Engine | F6L 912 | BF6L 913 | BF6L 913C | BF6L 913C | BF6L 913C |
| Cylinders diameter, mm | 180 | 200 | 200 | 200 | 200 |
| Cylinders stroke, mm | 1400 | 1400 | 1600 | 1800 | 2000 |
| Hopper capacity, l | 400 | 450 | 450 | 450 | 450 |
| Cycles per minute | 30 | 30 | 30 | 30 | 32 |
| Delivery, $m^3$/h | 64 | 79 | 90 | 101 | 120 |
| Gate pressure, MPa | 5,0 | 5,0 | 5,0 | 5,0 | 5,0 |
| Air compressor output, l/min. | 1050 | 1050 | 1050 | 1270 | 1270 |
| Air compressor pressure, MPa | 1,0 | 1,0 | 1,0 | 1,0 | 1,0 |
| Gate diameter, mm | 125 | 125...150 | 125...150 | 125...150 | 125...150 |
| Pump weight, kg | 6500 | 7000 | 7200 | 7800 | 8000 |
| Dimensions, m: | | | | | |
| L | 4,7 | 4,7 | 5,1 | 5,5 | 5,9 |
| B | 2,26 | 2,26 | 2,26 | 2,26 | 2,26 |
| H | 2,5 | 2,5 | 2,5 | 2,5 | 2,5 |
| h | 1,3 | 1,3 | 1,4 | 1,4 | 1,4 |

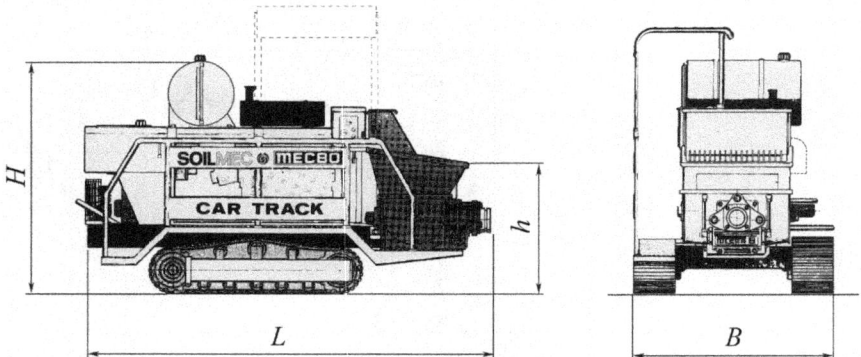

Figure A.5.17. SoilMec concrete pump

# TECHNICAL FEATURES
# OF CASAGRANDE ROTARY DRILLING RIGS

*Table A.6.1*

**Basic technical features of Casagrande rotary drilling rigs**

| Feature | Model | | | | | | | | |
|---|---|---|---|---|---|---|---|---|---|
| | B80 | B125 | B135 | B170 | B180HD | B250 | B300 | C600HD H40 |
| 1 | 2 | 3 | 4 | 5 | 6 | 7 | 8 | 9 |
| Maximum drilling depth by Kelly method, m | 38 | 50 | 50 | 50 | 68 | 68 | 70 | 87 |
| Maximum drilling diameter by Kelly method, m | 1300 | 1500 | 1500 | 1500 | 1800 | 2500 | 2000 | 2700 |
| Maximum drilling depth by continuous flight auger, m | 16 | 20.4 | 20.4 | 21.0 | 22.7 | 24.7 | 25.7 | 28.5 |
| Maximum drilling diameter by continuous flight auger, m | 600 | 800 | 800 | 800 | 1000 | 1200 | 1200 | 1200 |
| Width of extendable crawler track undercarriage, m | 2.5...3.6 | 2.5...3.7 | 2.5...3.7 | 2.5...3.9 | 2.5...3.9 | 3.0...4.4 | 3.0...4.4 | 3.0...4.4 |
| Track width, mm | 600 | 600 | 600 | 700 | 700 | 900 | 900 | 900 |
| Traction force of main winch, kN | 105 | 135 | 135 | 180 | 200 | 200 | 220 | 250 |
| Traction force of auxiliary winch, kN | 45 | 60 | 60 | 70 | 70 | 110 | 110 | 110 |

| 1 | 2 | 3 | 4 | 5 | 6 | 7 | 8 | 9 |
|---|---|---|---|---|---|---|---|---|
| Feed force of drilling tool when using hydraulic cylinder, kN: | | | | | | | | |
| upward | 140 | 148 | 148 | 148 | 211 | 211 | 211 | 400 |
| downward | 90 | 114 | 114 | 122 | 125 | 180 | 180 | 250 |
| Feed force of drilling tool when using winch, kN: | | | | | | | | |
| upward | 240 | 240 | 240 | 240 | 240 | 240 | 320 | 400 |
| downward | 240 | 240 | 240 | 240 | 240 | 240 | 320 | 400 |
| Maximum torsion torque at drilling tool, kN·m | 100 | 112 | 112 | 155 | 180 | 217 | 250 | 358 |
| Drilling rate, rpm | 42 | 33 | 33 | 32 | 34 | 34 | 29 | 25 |
| Unladen weight of rig, t | 27 | 35.5 | 35.5 | 52 | 62 | 80 | 88 | 100 |

# TECHNICAL FEATURES OF LIEBHERR MULTIFUNCTIONAL RIGS AND PILING AND VIBRO EQUIPMENT

*Table A.7.1*

**Technical features of Liebherr multifunctional rigs**

| Feature | Model | | |
|---|---|---|---|
| | LRB 125 | LRB 155 | LRB 255 |
| Mast height, m | 12.5 | 18; 21; 24 | 21; 24; 27 and 30 |
| Lifting capacity (hammer including driving cap of pile and pile), t | 12 | 15 | 30 |
| Maximum hammer weight, t | 6 | 8 | 15 |
| Maximum pile weight, t | 6 | 7 | 15 |
| Maximum traction force (mast on ground), t | 20 | 30 | 40 |
| Maximum torsion torque, kN·m | 120 | 220 | 300 |
| Swing radius from pile centerline to platform swing centerline, m | 3.3...5.5 | 3.0...4.7 | 3.2...4.9 |
| Maximum lean | | | |
| in transverse direction | ± 1 : 20 | ± 1 : 20 | ± 1 : 20 |
| forward in longitudinal direction | 1 : 6 | 1 : 6 | 1 : 6 |
| backward in longitudinal direction | 1 : 3 | 1 : 3 | 1 : 3 |
| Mast displacement: | | | |
| higher than ground level (depending on swing), m | 5 | 3 | 3 |
| lower than ground level (depending on mast length), m | 5 | 5 | 5 |
| Mast rotation, degree | ± 90 | ± 90 | ± 90 |
| Traction force of auxiliary winch (effective), kN | 50 | 80 | 80 |
| Downfeed force of drilling tool when using winch, kN | 150 | 300 | 450 |
| Upfeed force of drilling tool when using winch, kN | 200 | 300 | 450 |
| Engine capacity, kW | 450 | 450 | 670 |
| Traction force of undercarriage, kN | 437 | 632 | 622 |
| Width of extendable undercarriage, m | 3.0...4.2 | 3.2...4.5 | 3.5...4.7 |
| Undercarriage length on ground, m | 4.21 | 4.68 | 4.93 |

| Feature | Model | | |
|---|---|---|---|
| | LRB 125 | LRB 155 | LRB 255 |
| Track width of undercarriage, mm | 700 | 700, 800 and 900 | 900 |
| Maximum traveling speed, km/h | 2.3 | 1.5 | 1.5 |
| Transportation weight: | | | |
| of base machine (without mast, working attachments and counterweight) | 39.1* | 34.8 | 41.0 |
| mast 18,2 m long | – | 23.8 | - |
| mast 21,2 m long | – | 24.5 | 27.3 |
| mast 24,2 m long | – | 25.3 | 28.4 |
| mast 27,2 m long | – | – | 29.6 |
| mast 30,2 m long | – | – | 30.8 |
| counterweight | 3.9 | 8.0 | 12.5 |
| Unladen weight (base machine, minimum-length mast, counterweight), t | 43.0 | 66.6 | 80.8 |
| Soil bearing pressure, kgf/cm$^2$ | 0.83 | 0.79** | 0.91 |

\* To be transported together with mast.
\*\* At the track width of 900 mm.

*Table A.7.2*

## Technical features of Liebherr hydraulic hammers

| Feature | Type of hammer | | | | | |
|---|---|---|---|---|---|---|
| | H 50/3 | H 50/4 | H 85/5 | H 85/7 | H 110/7 | H 110/9 |
| Weight of striking part, t | 3.0 | 4.0 | 5.0 | 7.0 | 7.0 | 9.0 |
| Maximum impact energy, kN·m | 40 | 51 | 60 | 83 | 83 | 106 |
| Frequency of strikes per minute | 50...100 | 50...100 | 50...100 | 45...100 | 40...100 | 36...100 |
| Weight of hammer together with striking part, t | 6.2 | 7.2 | 8.6 | 10.5 | 10.6 | 12.7 |
| Length of driven element, m, using Liebherr rig: | | | | | | |
| LRB 125 | 12 | 12 | – | – | – | – |
| LRB 155 | – | – | 21 | 21 | – | – |
| LRB 255 | – | – | – | – | 27 | 27 |

## Technical features of Piling and Vibro Equipment high-frequency ring vibratory pile drivers

| Feature | Model | | |
|---|---|---|---|
| | 20 VMR | 32 VMR | 38 VMR |
| Static moment of eccentric weights, kgf·m | 0...20 | 0...32 | 0...38 |
| Maximum rotation speed, rpm | 2300 | 2300 | 2300 |
| Maximum applied force, kN | 1160 | 1800 | 1860 |
| Maximum pulling-out force, kN | 300 | 400 | 400 |
| Maximum pressing force, kN | 200 | 400 | 400 |
| Amplitude, mm | 0...6 | 0...5 | 0...5 |
| Casing pipe diameter, mm | 355...510 | 356...610 | 610...830 |
| Total weight, t | 6.9 | 12.0 | 12.5 |
| Lenght of driven pipe using Liebherr rig, m: | | | |
|     LRB 125 | 27 | – | – |
|     LRB 155 | 34 | – | – |
|     LRB 255 | - | 40 | 40 |

## Technical features of Piling and Vibro Equipment high-frequency ring vibratory pile drivers with clamping attachment to drive pipes, I beams and plank piles

| Feature | Model | |
|---|---|---|
| | 23 VML | 40 VML |
| Static moment of eccentric weights, kgf·m | 0...23 | 0...40 |
| Maximum rotation speed, rpm | 2300 | 2000 |
| Maximum centrifugal force, kN | 1350 | 1750 |
| Maximum pulling-out force, kN | 300 | 400 |
| Maximum pressing force, kN | 200 | 300 |
| Maximum amplitude (without clamp), mm | 17 | 19 |
| Total weight without clamp, kg | 3600 | 6200 |
| Total weight with clamp, kg | 5250 | |
| Dynamic weight without clamp, kg | 2700 | 4300 |
| Overall dimensions, mm: | | |
|     width | 1460 | 2570 |
|     height without clamp | 2095 | 2590 |
|     thickness | 795 | 740 |
| Maximum length of driven element using Liebherr rig, m: | | |
|     LRB 125 | 15.5 | – |
|     LRB 155 | 21.0 | – |
|     LRB 255 | – | 28.0 |

# TECHNICAL FEATURES OF FUNDEX MULTIFUNCTIONAL RIGS AND EQUIPMENT

*Table A.8.1*

## Technical features of Fundex multifunctional rigs

| Feature | Model | | | | | |
|---|---|---|---|---|---|---|
| | FN14 | F12SE | F2800 | F3500 | F4201 | F15 |
| Mast height, m | 18.0 | 21.0 | 28.0 | 35.0 | 24.6 (32.8) | 34.0 |
| Extended mast height, m | 24.0 | 34.0 (37.0) | 31.0 | 40.0 | 42.0 | 46.0 |
| Maximum torsion torque at drilling tool, kN·m | 100 | 450 | 450 | 450 | 400 | 400 |
| Feed force of drilling tool when using winch, kN: upward downward | 400 – | 800 350 | 800 400 | 1000 500 | 800 400 | 720 400 |
| Number of winches × traction force, kN | 1×100 | 3×100 | 3×100 | 3×125 | 3×100 | 1×100 |
| » | 1×50 | 1×50 | 1×50 | 2×50 | 1×50 | 2×100 |
| » | 1×40 | 1×40 | 1×15 | 1×9 | 2×15 | 1×60 |
| » | – | 1×9 | 1×9 | – | 1×9 | 1×9 |
| Track width, mm | 700 (850) | 700 (850) | 900 | 900 | 900 | 750 (850) |
| Width of extendable crawler track undercarriage, m | 3.0...4.5 | 3...4.5 | 3.5...4.9 | 3.5...5.0 | 3.5...5.0 | 3.5...6.0 |
| Transportability including attachments | Yes | In part | In part | In part | In part | No |
| Transportation weight, t | 38 | 48...50 | 60...62 | 72...74 | 90 | 86 |
| Number of hydraulic outriggers in front, ea. | Mast support | 2 | 2 | 2 | 2 | 2 |
| The same at rear, ea. | 2 | 2 | 2 | 2 | 2 | 2 |
| Possibility of U-turn | Yes | Yes | Yes | Yes | Yes | Yes |
| Possibility of driving sheet piling | Yes | Yes | Yes | Yes | Yes | Yes |
| Possibility of driving piles and pipes | Yes | Yes | Yes | Yes | Yes | Yes |
| Possibility of auger drilling | Yes | Yes | Yes | Yes | Yes | Yes |
| Possibility of borehole reaming | No | Yes | Yes | Yes | Yes | Yes |
| Hydraulic control | Yes | Yes | No | No | No | Yes |
| Electronic control and monitoring | On request | No | Yes | Yes | Yes | No |

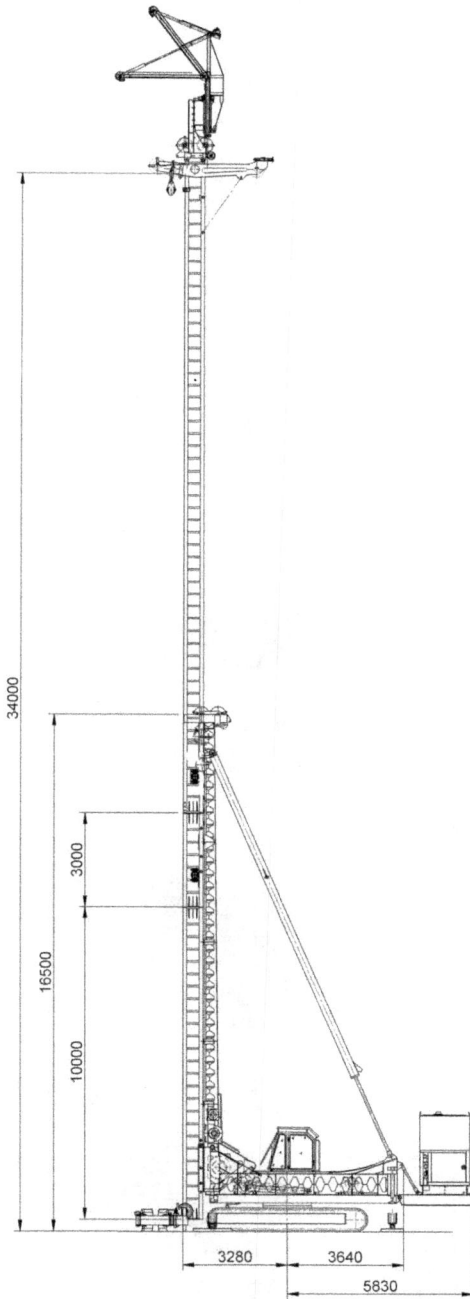

Figure A.8.1. Fundex F12SE multifunctional rig

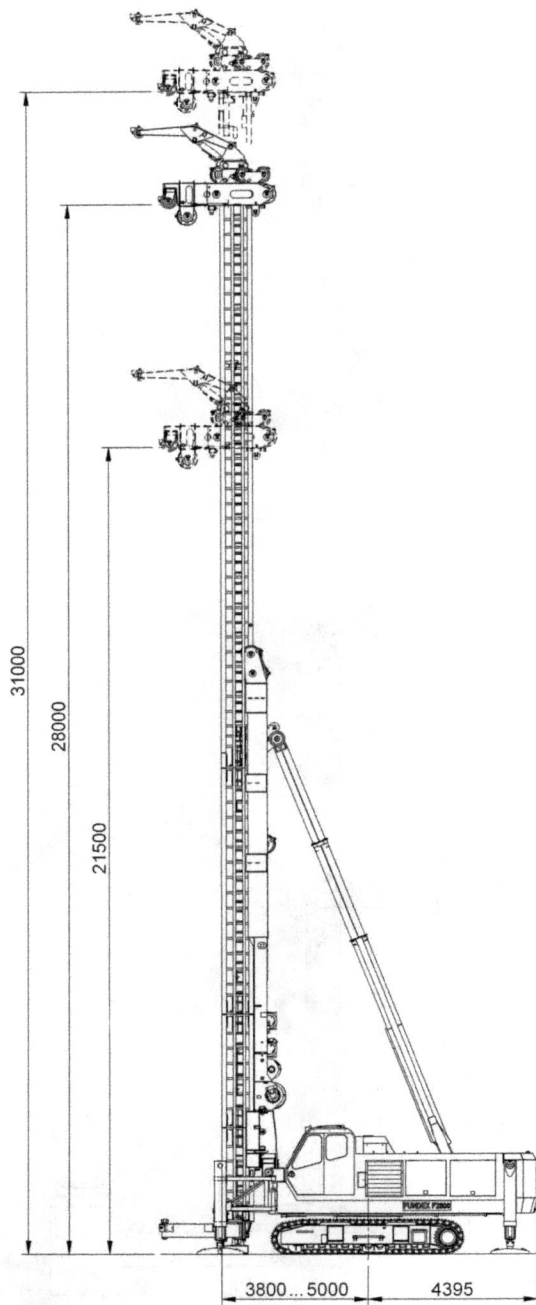

Figure A.8.2. Fundex F2800 multifunctional rig

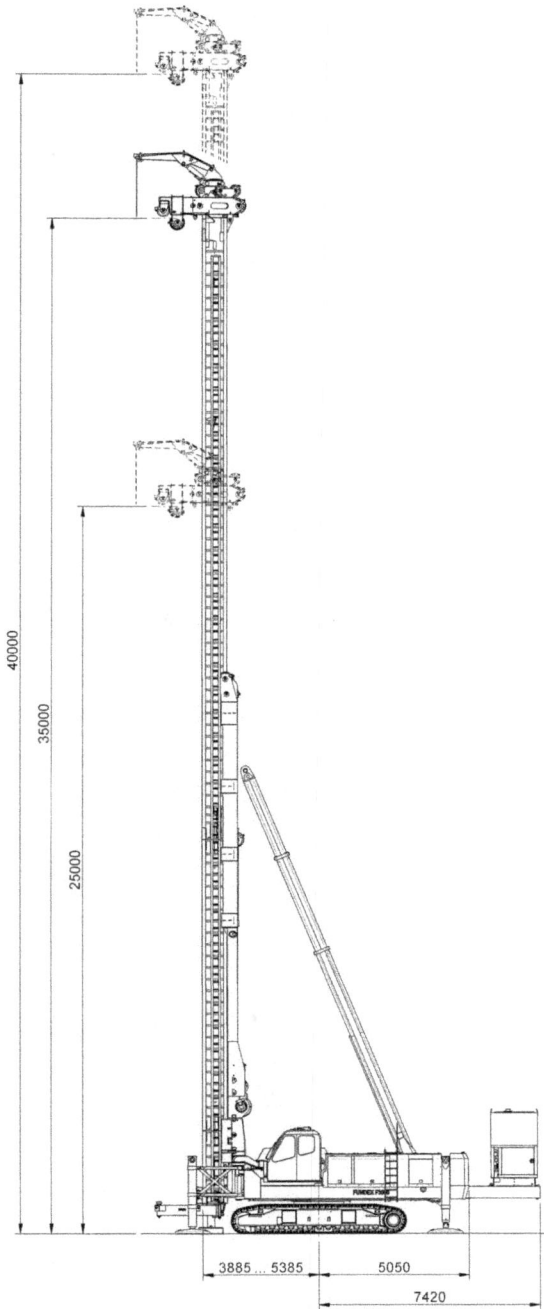

Figure A.8.3. Fundex F3500 multifunctional rig

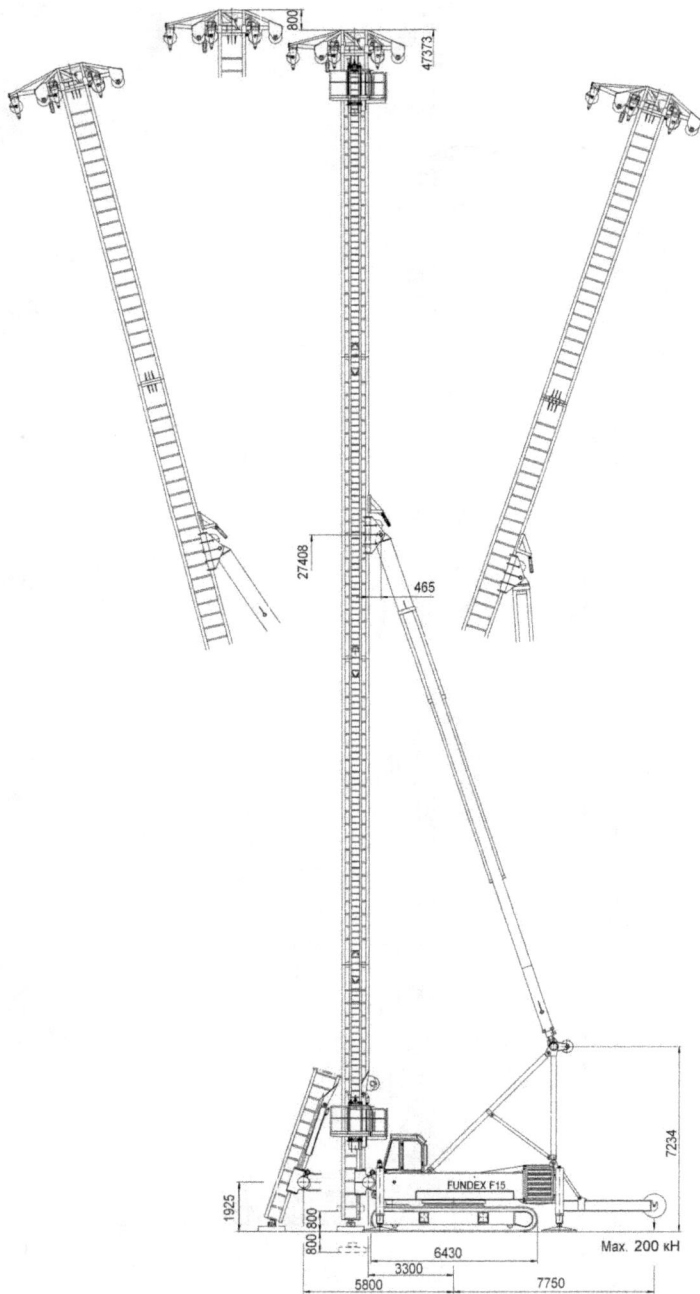

Figure A.8.4. Fundex F15 multifunctional rig

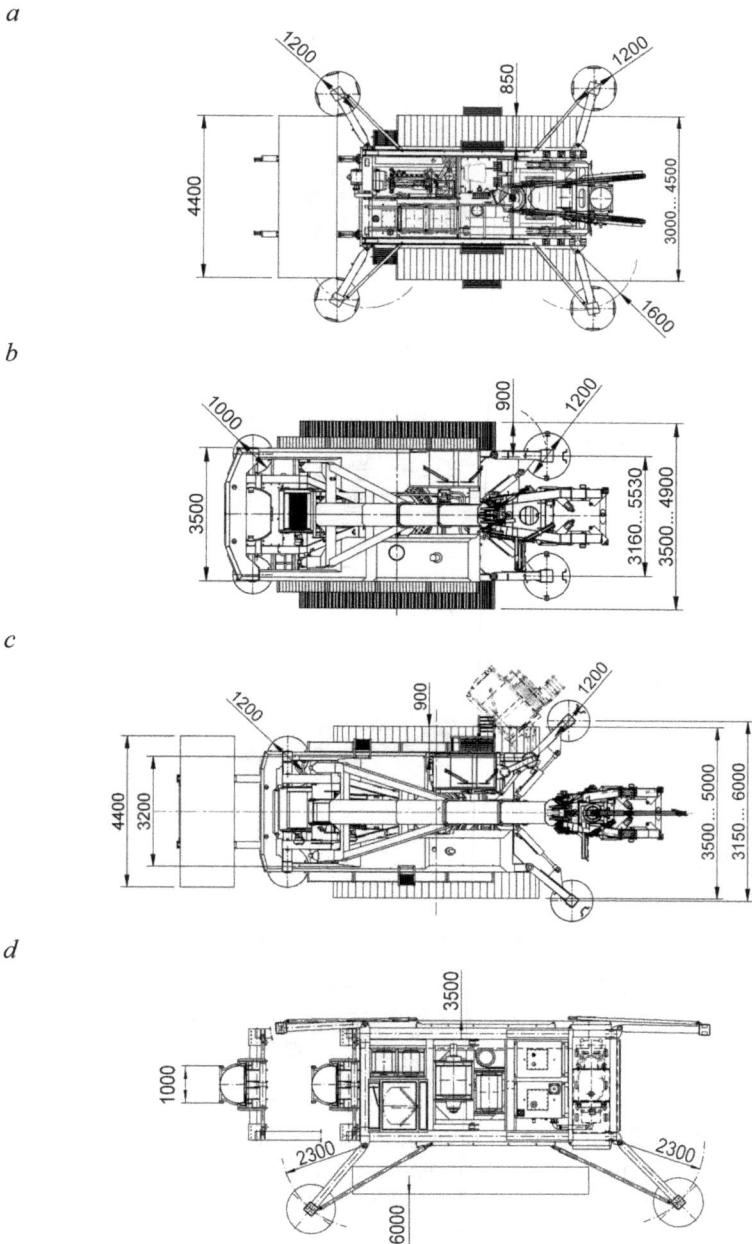

Figure A.8.5. Overall dimensions of multifunctional rigs
Fundex in plan view: *a* – Fundex F12SE; *b* – Fundex F2800;
*c* – Fundex F3500; *d* – Fundex F15

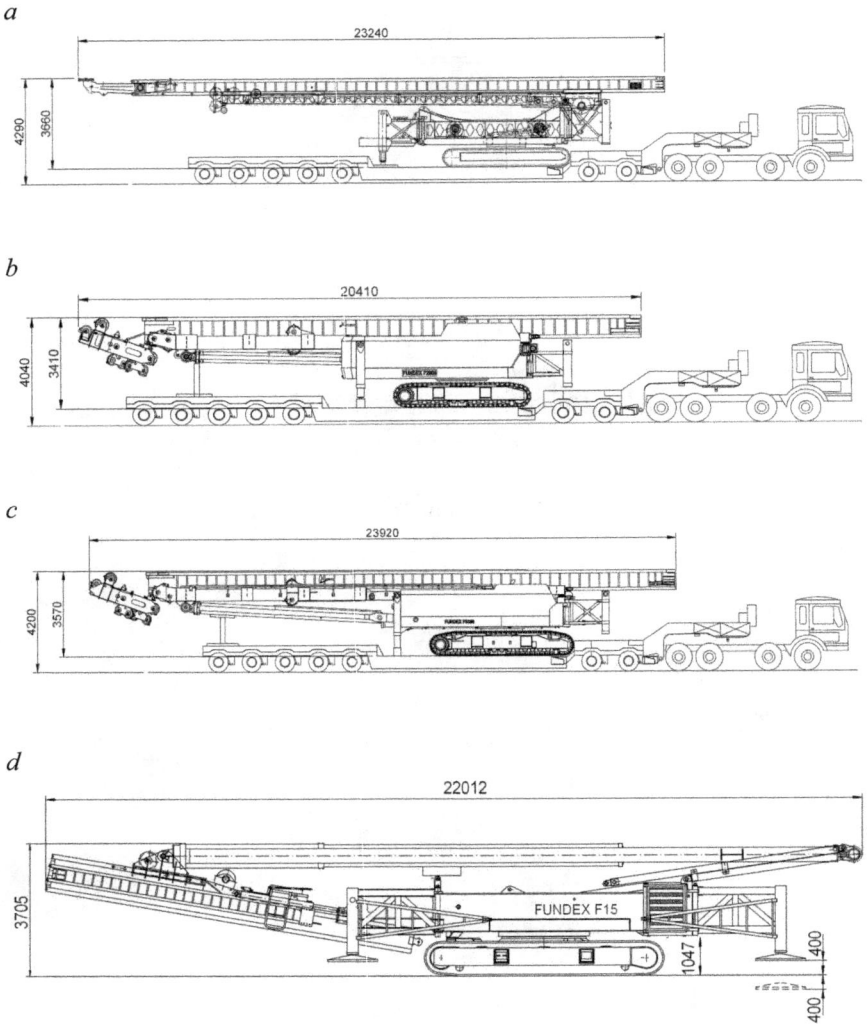

Figure A.8.6. Overall dimensions of Fundex multifunctional rigs
in transportation position: *a* – Fundex F12SE;
*b* – Fundex F2800; *c* – Fundex F3500; *d* – Fundex F15

## Technical features of Fundex ring vibrators

| Feature | Model of vibrator | | | |
|---|---|---|---|---|
| | FVE 40 | FVE 50 | FVE 50 HD | FVE 60 |
| Static moment of eccentric weights, kgf·m | 40 | 50 | 60 | 60 |
| Maximum rotation speed of eccentric weights, rpm | – | 1600 | 1600 | 1600 |
| Maximum driving orce, kN | 1000 | 1400 | 1600 | 1600 |
| Maximum linear force, kN | 400 | 800 | 1000 | 1000 |
| Maximum amplitude of vibrator, mm | 17 | 19 | 19 | 19 |
| Minimum pipe diameter, mm | 368 | 355 | 324 | 355 |
| Maximum pipe diameter, mm | 609 | 609 | 711 | 711 |
| Vibrating weight, kN | 60 | 60 | 60 | 61 |
| Total weight, kN | – | 64 | – | 65 |

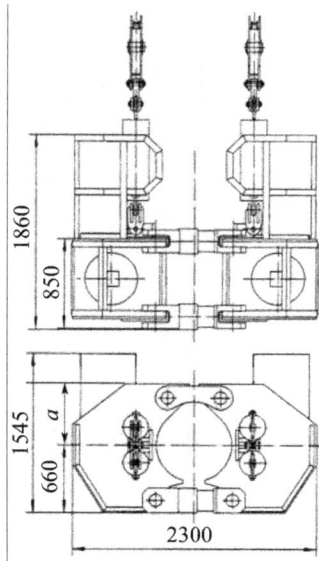

Figure A.8.7. Fundex ring vibrator
($a$ = 490 mm for FVE 50 vibrator, $a$ = 590 mm for FVE 60 vibrator)

## Technical features of Fundex vibratory hammers

| Feature | Make of hammer | | | | | |
|---|---|---|---|---|---|---|
| | 14RF | 18RF | 23RF | 28RF | 36RF | 46RF |
| Static moment of eccentric weights, kgf·m | 0...14 | 0...18 | 0...23 | 0...28 | 0...36 | 0...46 |
| Driving force, kN | 810 | 850 | 1334 | 1600 | 2030 | 2670 |
| Maximum rotation speed of eccentric weights, rpm | 2300 | 2300 | 2300 | 2300 | 2300 | 2300 |
| Maximum amplitude of vibrator, mm | 10 | 12 | 9 | 11 | 14 | 13 |
| Maximum static pulling force, kN | 240 | 240 | 400 | 500 | 500 | 800 |
| Maximum pressing force, kN | 240 | 240 | 400 | 500 | 500 | 800 |
| Maximum operating pressure, MPa | 34 | 34 | 34 | 34 | 34 | 34 |
| Unladen weight, kg | 4285 | 4295 | 6900 | 8000 | 9300 | 9750 |
| To be used together with Fundex rig * | F2800 | F2800 | F2800 | F3500 | F3500 | F3500 |

* It is possible to use with the other Fundex rigs.

# TECHNICAL FEATURES OF JUNTTAN RIGS AND EQUIPMENT

*Table A.9.1*

**Technical features of Junttan multifunctional rigs**

| Feature | Model | |
|---|---|---|
| | PM 26 | PM 28 |
| 1 | 2 | 3 |
| Make of mast | KE 269602 | KE 280101 |
| Length of mast sections, m | 4.0 + 13.0 + 8.0 | 12.0 + 10.0 + 4.4 |
| Mast height without jib, m | 25.0 | 26.4 |
| Additional extension of mast, m | 2.0, 4.0 | 5.0 |
| Lifting capacity of mast (weight of pile and hammer), t | 30 | 35 |
| Maximum mast displacement, mm: | | |
| vertically upward | 1700 | 1700 |
| vertically downward | 100 | 100 |
| horizontally left / right | 900 | 900 |
| Lifting capacity of main winch (for attached equipment: hammer or rotation head with drilling tool), t | 18 or 25 | 25 |
| Maximum engine capacity, kW | 414 | 414 |
| Dimensions of extendable crawler track undercarriage, mm: | | |
| length | 5700 | 5700 |
| minimum width | 3500 | 3500 |
| maximum width | 4780 | 4800 |
| track width | 900 | 900 |
| Distance from rig center of gravity to extreme point of counterweight, mm | 4700 | 5000 |
| Number of hydraulic outriggers in front, ea. | Mast support | Mast support |
| The same at rear, ea. | 2 | 2 |
| Overall dimensions of rig in transportation position, mm: | | |
| length | 26280 | 12100 |
| width | 3500 | 3500 |
| height without hammer / rotation head | 3350 | 3400 |
| Transportation weight, t | 65.0 | 65 |

| 1 | 2 | 3 |
|---|---|---|
| *Junttan with equipment to drive piles* | | |
| Rig height with mast without extension, mm | 26700 | – |
| Junttan hammers used | HHK 5A...HHK 12A, HHK 5S...HHK 9S | – |
| Lifting capacity of auxiliary winch (for pile), t | 10 or 12 | – |
| Weight of counterweight, t: | 4 | – |
| Unladen weight of rig, t | 77 | – |
| *Junttan with equipment for borehole percussion* | | |
| Rig height with mast without extension, mm | 27650 | – |
| Junttan hammers used | HHK 5A, HHK 6A, HHK 5S, HHK 6S | – |
| Lifting capacity of auxiliary winch, t | 10 or 12 | – |
| Weight of counterweight, t: | 10 | – |
| Unladen weight of rig (without shell for borehole percussion), t | 83 | – |
| *Junttan with equipment for borehole reaming* | | |
| Rig height with mast without extension, mm | 27650 | 28000 |
| Maximum drilling depth without mast extension, m | 22 | 22 |
| Maximum drilling depth with mast extension, m: | 26 | 32 |
| Reaming tip diameter, mm | 300...800 | 300...800 |
| Feed force of drilling tool when using winch, kN: upward downward | 1000 240 | 1000 240 |
| Maximum torsion torque at drilling tool, kN·m | 400 | 400 |
| Lifting capacity of auxiliary winch, t | 5 | 5 |
| Weight of counterweight, t: | 10 | 12 |
| Unladen weight of rig without reaming tool, t | 88 | 90 |

| 1 | 2 | 3 |
|---|---|---|
| *Junttan with equipment to drill boreholes by Kelly method* | | |
| Rig height with mast without extension, mm | 27730 | 28900 |
| Maximum drilling depth, m: | | |
| Kelly bar JKB 400 3/33 | 33 | 33 |
| Kelly bar JKB 400 3/41 | 41 | 41 |
| Maximum drilling diameter without casing pipe, mm: | | |
| rotation head JO 150 | 1500 | 1500 |
| rotation head JO 200 | 2000 | 2000 |
| Drilling diameter, mm | 750; 1000; 1200; 1500; 1700; 2000 | 750; 1000; 1200; 1500; 1700; 2000 |
| Feed force of drilling tool when using hydraulic cylinder, kN: | | |
| upward | 530 | 530 |
| downward | 450 | 450 |
| Maximum torsion torque at drilling tool, kN·m | 400 | 400 |
| Weight of counterweight, t: | 12 | 12 |
| Unladen weight of rig with JKB 400 3/41 bar and JO 200 rotation head, t | 115 | 120 |
| *Junttan with equipment for borehole drilling with continuous flight auger* | | |
| Rig height with mast without extension, mm | 27650 | 28820 |
| Maximum pile length without auger extension, m | 20 | 21 |
| Maximum drilling diameter, mm | 1200 | 1200 |
| Feed force of drilling tool when using winch, kN: | | |
| upward | 1000 | 1000 |
| downward | 240 | 240 |
| Maximum torsion torque at drilling tool, kN·m | 400 | 400 |
| Lifting capacity of auxiliary winch, t | 5 | 5 |
| Weight of counterweight, t: | 10 | 12 |
| Unladen weight of rig without auger, t | 88 | 86 |

*Table A.9.2*

## Technical features of Junttan pile-driving rigs

| Feature | Model | | | | | |
|---|---|---|---|---|---|---|
| | PM 20 | PM 20HLC | PM 20LC | PM 23 | PM 25H | PM 25LC |
| 1 | 2 | 3 | 4 | 5 | 6 | 7 |
| Make of mast | KE 200501 | KE 209501 | KE 209501 | KE 239902 | KE 259302 | KE 259302 |
| Length of mast sections, m | 13.8+3.0 | 13.8+4.2 | 13.8+4.2 | 12.6+4.8+1.8 | 13.8+6.0+1.8 | 13.8+6.0 |
| Additional extension of mast, m | 1.8+4.8 | 3.0 | 1.8 | 2.0 | 1.8..6.0 | 1.8..6.0 |
| Lifting capacity of mast (weight of pile and hammer), t | 13 | 18 | 16 | 14 | 20 | 16 |
| Maximum mast displacement, mm: | | | | | | |
| vertically upward | 1000 | 1000 | 1000 | 4700 | 1000 | 1000 |
| vertically downward | 500 | 500 | 500 | 4330 | 500 | 500 |
| horizontally left / right | 1500 | 1500 | 1500 | 6650 | 1500 | 1500 |
| Lifting capacity of auxiliary winch (for hammer), t | 10 | 12 | 10 | 10 | 15 | 10 |
| Lifting capacity of auxiliary winch (for pile), t | 8 | 10 | 10 | 8 | 10 | 10 |
| Maximum engine capacity, kW | 179 | 179 | 179 | 179 | 298 | 248 |
| Dimensions of extendable crawler track undercarriage, mm: | | | | | | |
| length | 4760 | 5050 | 5050 | 5760 | 5700 | 5750 |
| minimum width | 3150 | 3350 | 3350 | 3250 | 3400 | 3350 |
| maximum width | 4150 | 4350 | 4350 | 4250 | 4400 | 4350 |
| track width | 800 | 900 | 900 | 900 | 900 | 900 |
| Optional track width, mm | 900 | 800 | 800 | 800 | 1100 | 1100 |
| Distance from rig center of gravity to extreme point of counterweight, mm | 4100 | 4200...5700 | 4200...5700 | 4250...5700 | 4200...5700 | 4230...5730 |
| Weight of counterweight, t: | 5.2 | 6.2 | 6.2 | 6.9 | 5.9 | 5.9 |

| 1 | 2 | 3 | 4 | 5 | 6 | 7 |
|---|---|---|---|---|---|---|
| Maximum pile length, m, when using Junttan hammer: | | | | | | |
| HHK 3S | 16 (20) | | | 19 | | |
| HHK 4AL | 15 (19) | | | | | |
| HHK 4A | 14 (18) | | 15 (19) | 19 | | |
| HHK 4S | 14 (18) | | 15 (19) | 19 | | |
| HHK 5A | | 15 (20) | 15 (19) | 19 | 19 | 17 |
| HHK 5S | | 14 (19) | 14 (18) | | | 16 |
| HHK 6A | | 14 (19) | | | | |
| HHK 6S | | 14 (18) | | | | |
| HHK 7A | | | | | 18 | 16 |
| HHK 7S | | | | | 17 | 16 |
| HHK 9A | | | | | 16 | |
| Overall dimensions of rig in transportation position, mm: | | | | | | |
| length | 17700 | 18900 | 18900 | 20200 | 20700 | 20700 |
| width | 3150 | 3350 | 3350 | 3250 | 3400 | 33500 |
| height without hammer | 3550 | 3590 | 3550 | 3370 | 3650 | 3650 |
| height with hammer | 3700 | 3750 | 3700 | 3370 | 3800 | 3800 |
| Weight of rig, t: | | | | | | |
| transportation | 39 | 54 | 44 | 47 | 58 | 51 |
| unladen | 49...60 | 66...72 | 56...65 | 59...66 | 77...87 | 66...80 |

*Notes:* 1. In parenthesis, given are the maximum lengths of piles or pile sections when driving them by rigs with extended mast.
2. The unladen weight of a rig depends on the weight of a hammer used.

*Table A.9.3*

**Basic technical features of Junttan hydraulic hammers**

| Feature | Model | | | | | | | |
|---|---|---|---|---|---|---|---|---|
| | HHK 3A | HHK 4A | HHK 5A | HHK 7A | HHK 9A | HHK 12A | HHK 14A | HHK 16A |
| Maximum energy, kN·m | 35 | 47 | 59 | 82 | 106 | 141 | 164 | 188 |
| Drop height of hammer striking part, mm | 50...1200 | 50...1200 | 50...1200 | 50...1200 | 50...1200 | 50...1200 | 50...1200 | 50...1200 |
| Number of strikes per minute | 40...100 | 40...100 | 40...100 | 40...100 | 40...100 | 40...100 | 40...100 | 40...100 |
| Weight of hammer striking part, kg | 3000 | 4000 | 5000 | 7000 | 9000 | 12000 | 14000 | 16000 |
| Weight of hammer with driving cap for pipes, kg | 6200 | 7100 | 8600 | 11200 | 13400 | 20200 | 22500 | 25000 |
| Maximum cross section of driven element, mm: | | | | | | | | |
| metal pipes | Ø 600 | Ø 600 | Ø 750 | Ø 830 | Ø 830 | Ø 1030 | Ø 1030 | Ø 1030 |
| concrete piles | 370×370 | 420×420 | 470×470 | 520×520 | 520×520 | 670×670 | 670×670 | 670×670 |
| Hammer height, mm: | | | | | | | | |
| without driving cap | 4680 | 5050 | 5420 | 6160 | 6530 | 6500 | 7000 | 7500 |
| with driving cap | 5160 | 5530 | 5900 | 6640 | 7010 | 6980 | 7480 | 7980 |

| Feature | Model | | | | | | | |
|---|---|---|---|---|---|---|---|---|
| | HHK 18A | HHK 20A | HHK 3AL | HHK 4AL | HHK 5AL | HHK 5S | HHK 7S | |
| Maximum energy, kN·m | 212 | 235 | 24 | 31 | 39 | 74 | 103 | |
| Drop height of hammer striking part, mm | 50...1200 | 50...1200 | 50...800 | 50...800 | 50...800 | 50...1500 | 50...1500 | |
| Number of strikes per minute | 40...100 | 40...100 | 40...100 | 40...100 | 40...100 | 30...100 | 30...100 | |
| Weight of hammer striking part, kg | 18000 | 20000 | 3000 | 4000 | 5000 | 5000 | 7000 | |
| Weight of hammer with driving cap for pipes, kg | 28500 | 30500 | 5000 | 6200 | 7600 | 9000 | 12500 | |
| Maximum cross section of driven element, mm: | | | | | | | | |
| metal pipes | Ø 1320 | Ø 1320 | Ø 600 | Ø 600 | Ø 750 | Ø 750 | Ø 830 | |
| concrete piles | Ø 1320 | Ø 1320 | 400×400 | 400×400 | 450×450 | 450×450 | 550×550 | |
| Hammer height, mm: | | | | | | | | |
| without driving cap | 6420 | 6740 | 3885 | 4255 | 4625 | 5910 | 6650 | |
| with driving cap | 6970 | 7290 | 4365 | 4735 | 5105 | 6640 | 7380 | |

| Feature | Model | | | | | | | | |
|---|---|---|---|---|---|---|---|---|---|
| | HHK 9S | HHK 12S | HHK 14S | HHK 16S | HHK 18S | HHK 20S | HHK 25S |
| Maximum energy, kN·m | 132 | 176 | 206 | 235 | 265 | 294 | 368 |
| Drop height of hammer striking part, mm | 50...1500 | 50...1500 | 50...1500 | 50...1500 | 50...1500 | 50...1500 | 50...1500 |
| Number of strikes per minute | 30...100 | 30...100 | 30...100 | 30...100 | 30...100 | 30...100 | 30...100 |
| Weight of hammer striking part, kg | 9000 | 12000 | 14000 | 16000 | 18000 | 20000 | 25000 |
| Weight of hammer with driving cap for pipes, kg | 16000 | 22000 | 24700 | 27000 | 30700 | 33200 | 40000 |
| Maximum cross section of driven element, mm:<br>metal pipes<br>concrete piles | Ø 830<br>550×550 | Ø 1230<br>– | Ø 1230<br>– | Ø 1230<br>– | Ø 1320<br>– | Ø 1320<br>– | Ø1420<br>– |
| Hammer height, mm:<br>without driving cap<br>with driving cap | 7390<br>8120 | 7245<br>8190 | 7745<br>8690 | 8245<br>9190 | 7100<br>8070 | 7420<br>8390 | 7600<br>8490 |

# TECHNICAL FEATURES OF BANUT RIGS AND HYDRAULIC HAMMERS FOR PILE DRIVING

*Table A.10.1*

## Technical features of Banut hydraulic rigs

| Feature | Model | | | |
|---|---|---|---|---|
| | Banut 450 | Banut 555 | Banut 650 | Banut 655 |
| 1 | 2 | 3 | 4 | 5 |
| Recommended Banut hydraulic hammer | SuperRAM 4000 | SuperRAM 5000 | SuperRAM 6000 | SuperRAM 6000XL |
| Effective mast height, m | 11.0 | 15.0 | 18.6 | 15.0 |
| Maximum length of pile (pile section), m | 14.0 | 16.0 | 20.0 | 20.0 |
| Maximum mast displacement, mm: | | | | |
| higher than ground level | 3500 | 1000 | 1420 | 5500 |
| lower than ground level | 1500 | 1000 | 1180 | 1000 |
| Maximum lean, degree: | | | | |
| forward $\alpha_1$ | 18 | 18 | 18 | 18 |
| backward $\alpha_2$ | 45 | 45 | 45 | 45 |
| right $\beta_1$ | 12.5 | 14.0 | 18.0 | 18.0 |
| left $\beta_2$ | 12.5 | 14.0 | 18.0 | 18.0 |
| Weight of pile, t: | 5.5 | 6.0 | 6.0 | 8.5 |
| Torsion torque, kN·m | 80 | 100 | 130 | 180 |
| Lifting capacity of winch, t: | | | | |
| for pile | 5.5 | 10.0 | 10.0 | 12.0 |
| for hammer | 5.5 | 12.0 | 12.0 | 12.0 |
| Maximum height of rig $A$, mm | 20800 | 21250 | 26320 | 26500 |
| Distance from rig center of gravity to longitudinal axis of pile $B$, mm | 3200...4000 | 4000...5200 | 4095...5295 | 4450...5650 |

| 1 | 2 | 3 | 4 | 5 |
|---|---|---|---|---|
| Distance from rig center of gravity to extreme point of counterweight $C$, mm | 3500 | 4940 | 4290...5690 | 5100 |
| Overall dimensions of rig in transportation position, mm: | | | | |
| width $D$ | 3000 | 3100 | 3300 | 3300 |
| length $E$ | 18600 | 22370 | 18450 | 22000 |
| height $F$ | 3400 | 3300 | 3400 | 3400 |
| Track width, mm | 800 | 800 | 900 | 900 |
| Mast height $G$, mm | 16100 | 20000 | 24000 | 20000 |
| Width of mast cross section $K$, mm | 410 | 500 | 500 | 600 |
| Undercarriage | CAT 325 D | SR 40 T | SR 40 T | SR 40 T |
| Engine capacity, kW | 161 | 195 | 300 | 261 |
| Width of crawler track undercarriage $S$, mm: | 2200...3500 | 2300...3300 | 2400...3800 | 2400...3800 |
| Traction force of crawler track undercarriage, kN | 410 | 340 | 540 | 556 |
| Hydraulic retractable outriggers: | | | | |
| quantity, ea. | 2 | 2 | 2 | 2 |
| swing, mm | 800 | 800 | 800 | 700 |
| Swing of retractable counterweight, mm | – | – | 1400 | – |
| Weight of rig without auxiliary equipment, t | 39 | 55 | 55 | 65 |

*Notes:* 1. The effective mast height depends on the height of attached equipment used. The Table shows the effective mast height for the use of the recommended Banut hydraulic hammers. 2. The maximum angle of lean is only allowed when the outriggers are used. 3. A weight of a pile to be lifted to mast depends on the mast position and a weight of a hydraulic hammer.

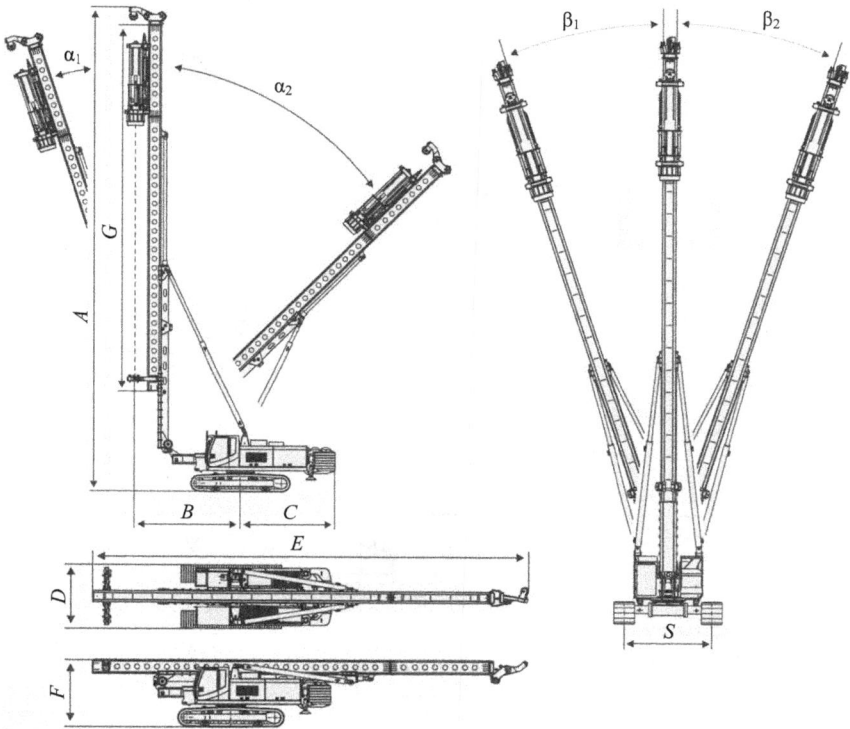

Figure A.10.1. Basic overall dimensions of Banut rig

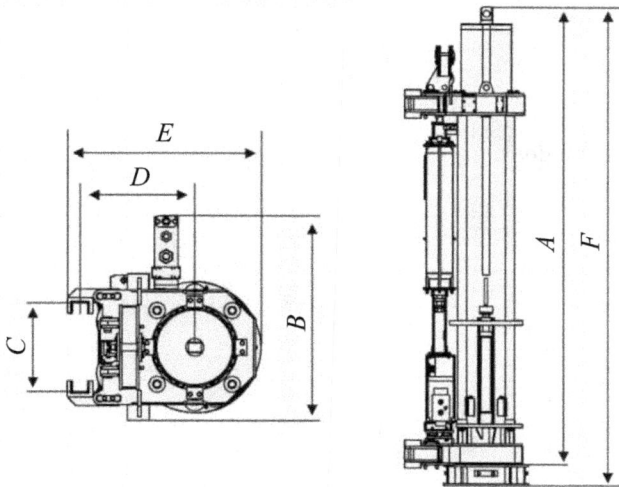

Figure A.10.2. Basic dimensions of Banut hydraulic hammer

## Technical features of Banut hydraulic hammers

| Feature | Model of hammer | | | | | | | |
|---|---|---|---|---|---|---|---|---|
| | SuperRAM 3000 | SuperRAM 4000 | SuperRAM 5000 | SuperRAM 6000 | SuperRAM 6000XL | SuperRAM 8000XL | SuperRAM 10000XL | SuperRAM 12000XL |
| Weight of hammer striking part, kg | 3000 | 4110 | 5060 | 6075 | 6110 | 8010 | 10020 | 12025 |
| Total weight of hammer without striking part, kg | 4900 | 6000 | 7000 | 8000 | 9200 | 11100 | 13100 | 15100 |
| Maximum drop height, mm | 1200 | 1200 | 1200 | 1200 | 1400 | 1400 | 1200 | 1200 |
| Number of strikes per minute | 100 | 100 | 100 | 100 | 100 | 100 | 100 | 100 |
| Maximum impact energy, kN·m | 35 | 47 | 58 | 70 | 82 | 109 | 117 | 141 |
| Required power, kW | 64 | 78 | 93 | 105 | 120 | 150 | 162 | 175 |
| Hammer height $A$, mm: in non-operating conditions | 3950 | 3865 | 3865 | 4120 | 4395 | 4395 | 4540 | 4665 |
| in operating conditions (maximum) | 5150 | 4680 | 4680 | 5320 | 5385 | 5525 | 5740 | 5865 |
| Hammer width $B$, mm | 1125 | 1125 | 1125 | 1125 | 1600 | 1600 | 1600 | 1600 |
| Width of guide $C$, mm | 500 | 500 | 500 | 500 | 500 | 500 | 500 | 500 |
| Distance from hammer centerline to center of guide $D$, mm | 630 | 630 | 630 | 630 | 715 | 715 | 715 | 715 |
| Depth $E$, mm | 1100 | 1100 | 1100 | 1100 | 1400 | 1400 | 1400 | 1400 |
| Hammer height with standard striking cap $F$, mm | 4665 | 4260 | 4260 | 4500 | 5030 | 5030 | 5180 | 5300 |

Förlag: BoD – Books on Demand, Stockholm, Sverige
Tryck: BoD – Books on Demand, Norderstedt, Tyskland

Rashid Aleksandrovitch **Mangushev**

Andrey Vladimirovitch **Ershov**

Anatoly Ivanovitch **Osokin**

# PILE CONSTRUCTION TECHNOLOGY

Second Revised and Updated Edition

Editor: *V.V. Surikova*

Computer-aided page proof by *D.A. Matveev*

Cover design by *A.V. Ershov*

Signed for printing 11.02.2015. Format 60×90 $^1/_{16}$.
Offset paper. Times type. Offset printing.
Conventional 14.25 printed sheets.

ASV Constraction, Sweden,
Mårdvägen 16  131 50 Saltsjö-Duvnäs

www.ingramcontent.com/pod-product-compliance
Lightning Source LLC
Chambersburg PA
CBHW060256220326
41598CB00027B/4123